John Edward Gray

Hand-list of the Edentate, Thick-Skinned and Ruminant Mammals in the

British Museum

John Edward Gray

Hand-list of the Edentate, Thick-Skinned and Ruminant Mammals in the British Museum

ISBN/EAN: 9783743378070

Manufactured in Europe, USA, Canada, Australia, Japa

Cover: Foto ©berggeist007 / pixelio.de

Manufactured and distributed by brebook publishing software (www.brebook.com)

John Edward Gray

Hand-list of the Edentate, Thick-Skinned and Ruminant Mammals in the

British Museum

HAND-LIST

OF THE

EDENTATE, THICK-SKINNED

AND

RUMINANT MAMMALS

IN THE

BRITISH MUSEUM.

BY

Dr. J. E. GRAY, F.R.S., F.L.S., &c.,

Keeper of the Zoological Department.

Forty-two Plates of Skulls, &c.

LONDON:

1873.

INTRODUCTION.

The List of Specimens of the Edentate, Thick-skinned and Ruminant Mammals and of their bones in the British Museum. These Specimens have been described—

1st. In the 'Catalogue of Specimens of Mammalia in the British Museum.' Part III.—Ungulata. Furcipeda. London, 1852. 12mo, with plates of skulls.

2nd. In the 'Catalogue of Carnivorous, Pachydermatous and Edentate Mammalia in the British Museum.' London, 1869. 8vo, with figures of skulls.

3rd. In the 'Catalogue of Ruminant Mammalia (*Pecora*, Linnæus) in the British Museum.' London, 1872. 8vo, with plates of skulls.

The new species and improvements in the arrangement of the animals that may have been discovered since the publication of these works have been adopted, and a reference made to where such species have been noticed, or are here described. When any of the specimens have served as the types of a description or figure of the species in any scientific work, reference is made to where the description and figure are to be found. Whenever the specimen has been presented or obtained directly from a collector, the name of the presentor or collector, and the habitat sent with the specimen, have been recorded immediately after the specimen; but the far greater number of the specimens have been obtained from dealers, with only the most general habitats. The condition in which the specimen is preserved, its size, and often the age or sex, when known with certainty, are recorded.

The following Table gives the number and state of the specimens at present in the Collection, and the manner in which they are preserved:—

Species.		Animals stuffed.	Skins unstuffed.	Skeletons.	Skulls.	Horns.	Tusks.	Teeth.	Animals in spirits.
49	Edentate Animals	37	54	49	64				24
69	Thick-skinned Animals	121	55	77	280	30	9	14	5
202	Ruminant Animals ...	436	179	135	587	465			1
320	Total	594	288	261	931	495	9	14	30

JOHN EDWARD GRAY.

British Museum,
     15th October, 1873.

TABLE OF CONTENTS.

HAND - LIST

OF

EDENTATE BEASTS

(BRUTA, *Linnæus*).

Order BRUTA.

Gray, Cat. Carniv. &c. p. 361.

Sub-Order 1. TARDIGRADA, *Gray, Cat. Carniv. &c.* p. 362.

Face short; body covered with dry harsh hair. Limbs elongate. Herbivorous.

Family 1. BRADYPODIDÆ, *Gray, Cat. Carniv. &c.* p. 362; *P. Z. S.* 1871, p. 428.

Tribe 1. CHOLŒPINA, *Gray, P. Z. S.* 1871, p. 431.

Hands two-clawed. The second and third fingers elongate, with well-developed bones; the first and fourth lateral fingers reduced to a single bone.—*Flower, Manual Osteol.* p. 274, f. 100. Feet three-clawed. Front grinders large, like the canines. Skull large. Brain cavity large; lower jaw produced in front. Neck short. Ribs 21—21, narrow, 8 lower on each side, floating. Lumbar vertebræ 5. Caudal vertebræ, &c., small, rudimentary.

1. CHOLŒPUS, *Gray, Cat. Carniv. &c.* p. 363; *P. Z. S.* 1871, p. 431.

Unau, *Cuv. Oss. Foss.* v. t. vi. and t. vii. f. 1 (skull and feet); t. iv. & t. v. (limbs). *Blainv. Osteogr.* t. i—iii (skeleton and skull). *Rapp, Edentata*, t. iii. f. 2.

1. CHOLŒPUS DIDACTYLUS, *Gray, Cat. Carniv. &c.* p. 363; *P. Z. S.* 1871, p. 431.

i. Animal, stuffed; half-grown.
S. America. 44, 2, 7, 10.

f. Animal, stuffed. Pale brown.
736 *b.* Skull of "*f.*" Young; nasal separate, developed. Like *Rapp, Edentata*, t. ii. f. 243. 46, 4, 25, 4.
S. America, Brazil. 44, 5, 14, 33.

c. Animal, stuffed; adult. Head dark brown. In store.
W. Indies.

B

d. Animal, stuffed ; adult. Head grey-brown.
    W. Indies. 42, 12, 27, 9.

c. Animal, stuffed. Dark reddish brown. Head pale.
    St. Juan de Fuca. Goodridge. 56, 10, 16, 234.
                           Presented by the **Haslar** Hospital.

b. Young, stuffed, from spirits ; bleached.
    Central America.

a. Animal, in spirits ; young.
    Central America.

h. Animal, in spirits ; 12 days old.
    South America. 63, 12, 19, 2.   Presented by Prof. Owen.

736 *d.* Skull; large, perfect.
    Ecuador. Parzudaki. 53, 6, 23, 29.

736 *c.* Teeth.
    S. America.

Var. Columbianus, Gray, P. Z. S. 1871, p. 430.

g. Animal, stuffed.
736 *a.* Skull of "*g.*" End cut off.
    Columbia. Parzudaki. 45, 9, 18, 12.

2. Cholœpus Hoffmanni, Gray, Cat. Carniv. &c. p. 363 ; P. Z. S.
1871, p. 432.

a. Animal, stuffed ; adult. Dark-coloured.
1510 *f.* Skull of "*a.*"
    Cordillera del Chuen. Salvin. 69, 7, 19, 11.

d. Animal, stuffed ; adult. Dark-coloured.
1510 *e.* Skull of "*d.*"
    S. America. 50, 1, 26, 4.   Presented by Capt. Kellett, R.N.

c. Animal, stuffed. Pale-coloured.
1510 *a.* Skull and bones of body of "*c.*"
    Costa Rica. 65, 3, 4, 5.   Presented by Dr. Peters.

b. Animal, stuffed ; young.
1510 *d.* Skull of "*b.*"
    Cholœpus Hoffmannii, *Gray, P. Z. S.* 1871, p. 432, f. 1 & 2. Lettered
      by mistake *C. didactylus.*
    Costa Rica. Salvin. 69, 7, 19, 14.

e. Skin, not stuffed.
1510 *c.* Skeleton of "*e.*"
    Costa Rica. Salvin. 69, 7, 19, 12.

f. Skin, not stuffed.
1510 *f.* Skull imperfect, of "*f.*"
    Costa Rica. Salvin. 69, 7, 19, 11.

g. Skin, not stuffed, with bones inside.
    Costa Rica. Carniol. 67, 8, 23, 1.

Tribe 2. BRADYPODINA, *Gray, P. Z. S.* 1871, p. 434.

Hands and feet three-clawed. The three middle digits nearly equally developed; the first and fifth metacarpals are present in a rudimentary condition, and have no phalanges.—*Flower, Manual Osteol.* p. 274. Neck elongate; cervical vertebræ 9. Ribs wide, 15—15, 3 lower on each side, floating. Lumbar vertebræ 4. Tail elongate, with 10 vertebræ. Skull short. Lower jaw truncated in front.

Ai, *Cuvier Oss. Foss.* v. t. iv. v. (skeleton). *Blainv. Osteogr.* t. ii. (skeleton) t. iv. (bones of trunk), t. vi. (limbs), *Gray, P. Z. S.* 1849, p. 66, fig. (skull), 1871, p. 435 (skulls).

2. BRADYPUS, *Gray, Cat. Carniv. &c.* p. 363; *P. Z. S.* 1871, p. 435.

   1. BRADYPUS CRINITUS, *Gray, Cat. Carniv. &c.* p. 364; *P. Z. S.* 1871, p. 435.

56 *a.* Animal, stuffed.
923 *b.* Skull of "*a.*"
   Brazils, Rio Janeiro.   Presented by Lord Stuart de Rothsay.

56 *b.* Animal, stuffed.
   S. America. 55, 10, 16, 234.   Presented by the Haslar Hospital.

56 *c.* Animal, stuffed.
   S. America.

923 *a.* Skull and skeleton.
   B. crinitus, *Gray, P. Z. S.* 1871, t. x. f. 1 (skull).
   Brazils. Becker. 47, 4, 6, 5.

   2. BRADYPUS AFFINIS, *Gray, Cat. Carniv. &c.* p. 364; *P. Z. S.* 1871, p. 436.

737 *a.* Skeleton with skull.
   Bradypus affinis, *Gray, P. Z. S.* 1849, p. 68, t. x. f. 2 (skull).
   Brazils. Brandt.

737 *b.* Skull.
   Bradypus affinis, *Gray, P. Z. S.* 1849, p. 68, t. x. f. 2.
   S. America.

3. ARCTOPITHECUS, *Gray, Cat. Carniv. &c.* p. 364; *P. Z. S.* 1871, p. 436.

   1. ARCTOPITHECUS CUCULLIGER, *Gray, P. Z. S.* 1871, p. 440.

55 *b.* Animal, stuffed. Male.
   Bradypus tridactylus, *var.*, *Gray, Cat.* 1843, p. 193.
   S. America.

d. Animal, stuffed. Male.
921 *c.* Skull of "*d*," 2¼ in.
   Demerara. 55, 8, 26, 6.

c. Animal, stuffed; young.
   A. cuculliger, *Gray, P. Z. S.* 1871, p. 441.
   S. America. Warwick. 49, 4, 26, 2.

a. Animal, stuffed. Female.
Bradypus tridactylus.
S. America.

921 *b*. Skeleton; young.

    2. ARCTOPITHECUS GULARIS, *Gray, Cat. Carniv. &c.* p. 364.; *P. Z. S.* 1871, p. 441.

a. Animal, stuffed.
Surinam.   Bartlett.   71, 5, 19, 12.

    3. ARCTOPITHECUS BLAINVILLEI, *Gray, Cat. Carniv. &c.* p. 365. *P. Z. S.* 1840, p. 71.

55 *c*. Animal, stuffed. Male.
919 *h*. Skull imperfect, of "*c*."
Upper Ucayale, Amazons.   Bartlett.   66, 3, 28, 13.

d. Animal, unstuffed. Female. No skull.
Upper Ucayale, Amazons.   Bartlett.   66, 3, 28, 12.

919 *g*. Skull of young. Imperfect.
Upper Ucayale, Amazons.   Bartlett.

55 *a*. Animal, stuffed. Female.
Bradypus tridactylus, "*a*," *List of Mam. in B. M.*, 1843, p. 193.
S. America.

55 *b*. Animal, stuffed. Female.
S. America.   Zool. Soc.

55 *c*. Animal, stuffed. Female.
S. America.   44, 7, 4, 14.

919 *a*. Skeleton and skull separate.
Arctopithecus Blainvillei, *Gray, P. Z. S.* 1840, p. 72, t. xi. f. 2 (skull).
Brazils.   Becker.

919 *d*. Skeleton.
Guiana.   Brandt.

919 *e*. Skeleton, very young.
Zool. Soc.

919 *f*. Skull, back imperfect.
S. America, Guiana?   67, 4, 12, 299.

919 *b*. Skull.
S. America.   Presented by the Zool. Soc.

919 *c*. Skull.
S. America.   49, 5, 5, 3.   Presented by the Zool. Soc.

    4. ARCTOPITHECUS BOLIVIENSIS, *Gray, P. Z. S.* 1871, p. 442.

a. Animal, stuffed. Male.
921 *b*. Skull of "*a*."
Arctopithecus gularis, *Gray, P. Z. S.* 1840, t. ix. f. 6.
Bolivia.   Bridges.   46, 7, 28, 24.

b. Animal, unstuffed; young.
920 *c.* Skull of "*b.*"
   Arctopithecus marmoratus, *Gray, P. Z. S.* 1849, t. ix. f. 4 (lower
     jaw).
   A. boliviensis, *Gray, P. Z. S.* 1871, p. 442, f. 3 & 4 (skull).
   S. America. 42, 4, 29, 19.

921 *b.* Skeleton and skull, half-grown.
   Bolivia? Bridges. 46, 10, 16, 14.

920 *a.* Skull, young.
   Bolivia. Bridges.

   5. ARCTOPITHECUS MARMORATUS, *Gray, Cat. Carniv. &c.* p. 364;
     P. Z. S. 1871, p. 443.

a. Animal, stuffed. Female; adult.
920 *a.* Skull of "*a.*" 49, 4, 13, 3.
   A. marmoratus, *Gray, P. Z. S.* 1849, p. 71, t. xi. f. 3 (lower jaw).
   S. America. 42, 4, 29, 20.

b. Animal, stuffed; female. Fur between shoulders soft.
   A. marmoratus, var., *Gray, P. Z. S.* 1871, p. 443.
   S. America. 55, 10, 16, 233.   Presented by the Haslar Hospital.

c. Animal, stuffed; young.
   S. America.

   6. ARCTOPITHECUS CASTANEICEPS, *Gray, P. Z. S.* 1871, p. 444.

a. Animal, stuffed. Male.
   A. castaneiceps, *P. Z. S.* 1871, p. 445, t. xxxv.
738 *a.* Skull of "*a.*"
   A. castaneiceps, *Gray, P. Z. S.* 1871, p. 445, f. 5.
   S. America, Nicaragua. Seeman. 71, 4, 8, 2.

   7. ARCTOPITHECUS GRISEUS, *Gray, P. Z. S.* 1871, p. 446.

a, b, c, d. Animals, stuffed; adult, male; female adult and two
     young.
   A. griseus, *Gray, P. Z. S.* 1871, p. 446, t. xxxvi.
1549 *c.* Skull of "*c.*"
   A. griseus, *Gray, P. Z. S.* 1871, p. 446, fig. 6.
1549 *a.* Skull of "*a.*"
1549 *b.* Skull of adult, imperfect. Of female "*b.*"
   Costa Rica, Veragua. Salvin. 69, 7, 19, 7, 8, 9, 10.

   8. ARCTOPITHECUS FLACCIDUS, *Gray, Cat. Carniv. &c.* p. 365;
     P. Z. S. 1871, p. 448.

a. Animal, stuffed. Male.
   A. flaccidus, *Gray, P. Z. S.* 1871, p. 448, t. xxxvii.
   A. flaccidus, var. 1. Dysonii, *Gray, P. Z. S.* 1849, p. 72.
922 *a.* Skull, adult; imperfect, of "*a.*"
   A. flaccidus, *Gray, P. Z. S.* 1849, p. 72, t. xi. f. 1 (skull).
   Venezuela. Dyson. 49, 4, 13, 3.

b. Animal, stuffed ; young.
   A. flaccidus, *var.* 2, Smithii, *Gray, P. Z. S.* 1849, p. 72.
922 *b.* Skull of " *b.* "
   Para.   Presented by J. P. G. Smith, Esq.
738 *b.* Skull, adult, complete ; and imperfect skeleton.
   A. problematicus, *Gray, P. Z. S.* 1849, p. 73. t. xi. f. 5 (lower jaw).
   B. problematicus, *Gerrard, Cat. Bones,* p. 290.
   Para.   44, 10, 9, 34.   Presented by J. P. G. Smith, Esq.
722 *c.* Skeleton, with skull.
   Guiana.   Brandt.   52, 9, 20, 5.

Sub-Order 2.   EFFODIENTIA, *Gray, Cat.* p. 361.

Face and tongue elongate.   Limbs short.   Pelvis simple.
Insectivorous.

Section 1. CATAPHRACTA, *Gray, Cat. Carniv. &c.* p. 366.

Family 1. MANIDÆ, *Gray, Cat. Carniv. &c.* p. 366.

Tribe 1. MANINA.

Front feet and fore arms hairy, without scales.   Tail attenuated,
   much longer than the body and head.   The central dorsal
   series of scales continued to the end of the tail.   Ears without
   any external couch.   Climbing trees?

Dr. Fitzinger has compiled a monograph of this family.   (See
' Sitzungsbericht,' 1872, p. 9), but, as usual, he only knows the species
from books, and makes several nominal species, which cannot be
distinguished in Museums.

1. MANIS, *Gray, Cat. Carniv. &c.* p. 367.

Dorsal scales large, lax.   Lateral ones elongate, keeled, in 11 (5—
1—5) longitudinal series.   Caudal in five series, 2—1—2.   Fore arms
naked, hairy.   The middle front toe very large ; lateral one smaller.

1. MANIS LONGICAUDA, *Gray, Cat. Carniv. &c.* p. 367.
118 *a.* Animal, stuffed.   38 in.
   Manis tetradactyla, *Gray, Cat. Mam.* p. 188.
   W. Africa, Guinea.   Called *Attadillo.*
118 *c.* Animal, stuffed, 39 in.
   W. Africa.
118 *b.* Skin, flat on board.
   W. Africa.
118 *d.* Skin, unstuffed ; 45 in.
1483 *a.* Skeleton of " *d.* "
   W. Africa.   Gaboon.   64, 12, 1, 10.
118 *e.* Animal, in spirits.
   W. Africa.   67, 5, 31, 3.   Presented by A. Swanzy, Esq.

2. PHATAGIN.

**Manis ** **; Phatagin, *Gray, Cat. Carniv. &c.* p. 368.

Dorsal scales small, keeled, tricuspidate, in 21 (10—1—10) longitudinal series of scales. Caudal scales in 5 series; marginal scales 35 to 37. Fore arms and hands naked, hairy. The middle front toe very large, the rest smaller.

1. PHATAGIN TRICUSPIS.

Manis tricuspis, *Gray, Cat. Carniv. &c.* p. 368.
Triglochinopholis tricuspis, T. multiscutata, T. tridentata and Pholidotus Gonyi, *Fitz. l. c.* p. 27—37.

115 *b.* Animal, stuffed, 17 in.
Manis multiscutata, *Gray, P. Z. S.* 1843, p. 22.
W. Africa.

115 *c.* Animal, stuffed, 30 in.
Skull of " *c*," *Gray, P. Z. S.* p. 369, f. 42.
W. Africa. W. Rich. 43, 2, 23, 2. Called *Galilali*. 46, 5, 13, 18.

115 *d.* Animal, stuffed, on stump, 24 in.
Manis multiscutata, *Zool. Soc. Cat.*
W. Africa. 55, 12, 24, 128, 9.

115 *e.* Animal, stuffed, 18 in.
W. Africa. 51, 8, 26, 1.

115 *h.* Skin, not stuffed, 21 in.
Africa. Zool. Soc. 52, 9, 18, 2.

115 *l.* Flat skin, 34 in.
Manis tridentata. Toulon.
W. Africa. Gaboon. 51, 11, 5, 3.

185 *i.* Skin, not stuffed.
S. Africa. 59, 5, 72, 21.   Presented by Sir A. Smith.

115 *j.* Animal, in spirits.
W. Africa. 48, 7, 11, 5.

115 *k.* Animal, in spirits.
Fernando Po. Dalton. 65, 3, 30, 9.

727 *b.* Skeleton.
Manis tricuspis, *Gray, l. c.* p. 369; f. 43. (Skull).
W. Africa. Du Chaillu. 63, 2, 13, 23.

Tribe 2. PHOLIDOTINA.

Fore arms and hands covered with scales. Tail attenuated, not longer than the head and body. The central dorsal series of scales continued to the end of the tail.

3. PHOLIDOTUS, *Gray, Cat. Carniv. &c.* p. 370.

Back with 17 longitudinal series of scales. The lateral ventral scales elongate; strongly keeled. Caudal shields in 5 series (1—3—1) above, and 7 series (1—5—1) at the base beneath. Keeled scales on the margin, 24—28 on each side.

Ears surrounded by a raised rim.
Burrowing.

1. PHOLIDOTUS JAVANUS.

Manis aspera, *Rapp, Edentata*, t. ii. *a*, t. vi. f. 1 & 2 (skull).

29 *b*. Animal, stuffed; male, 32 in. Tail with 24 marginal scales.
Java. 37, 4, 28, 72.

29 *c*. Animal, stuffed; female, 30 in. Tail with 24 marginal scales.
Java.

29 *d*. Animal, stuffed, 36 in. Male. Tail with 27 marginal scales.
Java. 41, 1, 8, 12.

29 *e*. Animal, stuffed, 33 in. Tail with 27 or 28 marginal scales.
Singapore. 42, 4, 12, 10.

29 *a*. Flat skin on board, 21 in. Tail imperfect.
Java.

29 *b*. Skull, 4½ in.
Java? Mus. Utrecht. 67, 4, 12, 298.

4. PANGOLIN.

Pholidotus ※, *Gray, Cat. Carniv. &c.* p. 371.

Back with 15 (7—1—7) or 13 (6—1—6) longitudial series of scales; the lateral series of scales broad, keeled. Caudal shields in 5 (1—3—1) longitudinal series above and below to the base. Keeled scales on the margin on each side, 17 or 18. Ears with a produced lobe on the edge.

Burrows, and enters the water! Swinhoe.

※ *Dorsal scales dark, in* 15 (7—1—7) *longitudinal series.*

1. PANGOLIN DALMANNII.

Pholidotus Dalmannii, *Gray, Cat. Carniv. &c.* p. 371; *Swinhoe, P. Z. S.* 1870, p. 237 & 630.

a. Animal, stuffed, 34½ in. Front claws very large, white.
China.

b. Animal, unstuffed, 39 in. Male. Front claws very large, white.
Pholidotus Dalmannii, *Swinhoe, P. Z. S.*
S.W. Formosa. 72, 11, 13, 2. Presented by R. Swinhoe, Esq.

c. Animal, unstuffed, 29 in. Female. Front claws very large, white.
S.W. Formosa. 72, 11, 13, 3. Presented by R. Swinhoe, Esq.

d. Flat skin, 33 in. Female. Front claws moderate.
Hainan. 72, 11, 13, 1. Presented by R. Swinhoe, Esq.

e. Animal, unstuffed, 30 in.   Front claws moderate.
  Formosa.   72, 11, 13, 3.   Presented by R. Swinhoe, Esq.

f. Animal, stuffed, 24 in.   Tail rather short.   Front claws large,
  white.
  Pholidotus Dalmannii, *Swinhoe*, **P. Z. S.**
  Amoy.   72, 11, 13, 5.   Presented by R. Swinhoe, Esq.

g. Animal, stuffed, 20 in.   Tail very short ; only 17 marginal scales.
  Pholidotus Dalmannii, *Swinhoe*, **P. Z. S.**
  Amoy.   72, 11, 13, 6.   Presented by R. Swinhoe, Esq.

h. Animal, stuffed, 17½ in.   Female.   Tail very short ; only 15 mar-
  ginal scales.
  Pholidotus Dalmannii, *Swinhoe*, **P. Z. S.**
  Amoy.   72, 11, 13, 7.   Presented by R. Swinhoe, Esq.

i. Flat skin, 16 in. ; young.   Tail moderate ; 18 or 19 marginal scales.
  Hainan.   72, 11, 13, 8.   Presented by R. Swinhoe, Esq.

j. Bones of body of adult ; female.
  Formosa.   72, 11, 13, 9.   Presented by R. Swinhoe, Esq.

k. Bones of body of young female.
  Formosa.   72, 11, 13, 9.   Presented by R. Swinhoe, Esq.

l. Bones of body of young.
  Amoy.   72, 11, 13, 9.   Presented by R. Swinhoe, Esq.

m. Bones of body.
  Formosa.   72, 11, 13, 9.   Presented by R. Swinhoe, Esq.

n. Skull, 3½ in. ; complete, with cartilage.
  Formosa. Swinhoe.   72, 11, 13, 10.   Presented by R. Swinhoe, Esq.

o. Skull, 3⅜ in.   Female.
  Amoy.   72, 11, 13, 10.   Presented by R. Swinhoe, Esq.

p. Skull, 3½ in.   Male.
  Amoy.   72, 11, 13, 10.   Presented by R. Swinhoe, Esq.

q. Skull, 2¾ in.   Female.
  Amoy.   72, 11, 13, 10.   Presented by R. Swinhoe, Esq.

r. Skull, 2⅝ in.   Male ; young.
  Amoy.   72, 11, 13, 10.   Presented by R. Swinhoe, Esq.

1* PANGOLIN, n. s.

s. Skull, 2¾ in. ; much shorter and thicker ; palate broader, imperfect.
  S.W. Formosa.   72, 11, 13, 10.   Presented by R. Swinhoe, Esq.

2. PANGOLIN GIGANTEUS.
Pholidotus giganteus, *Gray, Cat. Carniv. &c.* p. 373.

a. Animal, stuffed, 5 feet 10 in.
1458 *b.* Skull of "*a*," 5¾ in.
  Cape Coast Castle.

b. Skin, not stuffed, 39 in.   Tail with 17 marginal scales.   Feet not
  in good state.
  W. Africa.

c

c. Animal, stuffed, 32 in.   Tail with 17 marginal scales.
P. Africanus, *Gray, P. Z. S.* 1865, p. 369, fig.
W. Africa. Baikie.   65, 3, 30, 3.

1458 *a.* Skeleton, mounted, 54 in.
Pholidotus giganteus, *Gray, Cat. Carniv. &c.* p. 374, f. 44 (skull).
P. Africanus, *Gray, P. Z. S.* 1865, p. 368, t. xvii.
W. Africa, Gaboon.   Du Chaillu.

** *Dorsal shields pale whitish, in* 13 (6—1—6) *longitudinal series.*

3. PANGOLIN INDICUS.
Pholidotus indicus, *Gray, Cat. Carniv. &c.* p. 373.
Pangolin à courte queue, *Cuv. Oss. Foss.* v. t. viii. (skeleton).
Phatages laticaudatus and P. bengalensis, *Fitz. l. c.* p. 67—72.
Pholidotus auritus, P. Assamensis, and P. leucurus, *Fitz.* p. 63, 57
    & 52.

116 *g.* Animal, stuffed, 43 in.
India.   41, 94.

116 *a.* Young, stuffed.
Manis pentadactyla, "*a*," *Gray, Cat. Mam. B. M.,* 1843, p. 188.
India.

116 *d.* Skin, not stuffed, — in.
India.   Nepaul.   48, 6, 11, 7.   Presented by B. H. Hodgson, Esq.

116 *c.* Animal, stuffed, 34 in.
India.   45, 14, 29, 15.

116 *b.* Animal, stuffed, 27 in.
India.   37, 6, 10, 1.   Presented by J. Reid, Esq.

116 *f.* Flat skin, 34 in.
Manis auritus, *Hodgson.*
India.   58, 6, 24, 35.   Presented by B. H. Hodgson, Esq.

116 *e.* Flat skin, 37 in.
Manis auritus, *Hodgson.*
India.   58, 6, 24, 34,   Presented by B. H. Hodgson, Esq.

116 *h.* Animal, unstuffed, 29 in.
India.   66, 7, 20, 1.   Presented by Dr. Falconer.

116 *i.* Animal, unstuffed, 30 in.   Bad state.
India.   66, 7, 20, 2.   Presented by Dr. Falconer.

116 *j.* Animal, in spirits.   Ears well developed.
Nepaul.   Presented by B. H. Hodgson, Esq.

728 *i.* Bones of body.
Nepaul.   58, 6, 24, 158.   Presented by B. H. Hodgson, Esq.

728 *a.* Bones of body.
Manis auritus, *Hodgson.*
Nepaul.   45, 1, 12, 462.   Presented by B. H. Hodgson, Esq.

728 *b.* Skull, 3¾ in.   Rather imperfect.
India.

728 *c.* Skull, 3¾ in.
Darjeeling.   Presented by Prof. Oldham.

728 *d.* Bones of body and skull, 3½ in.
Nepaul. 58, 6, 21, 4. Presented by B. H. Hodgson, Esq.

728 *e.* Bones of body.
Nepaul. Presented by B. H. Hodgson, Esq.

728 *f.* Bones of body.
Nepaul. Presented by B. H. Hodgson, Esq.

728 *g.* Skull, 3¾ in.
India.

Tribe 3. SMUTSIANA. Tail very broad, shorter than the body; central series of shields only continued on the base of the tail. Ears with only a raised edge.

5. SMUTSIA, *Gray, Cat. Carniv. &c.* p. 374.

Ears without any external couch. Scales of the body in 11 series, upper ones broad, the lateral series very long. Tail with two series of shields on each side, lateral series very large. Feet scaly to the toes.

1. SMUTSIA TEMMINCKII, *Gray, Cat. Carniv. &c.* p. 375.
Phatages giganteus and P. Hedenborgii, *Fitzinger*, pp. 75—77.

a. Animal, stuffed, 36 in.
E. Africa? 44, 10, 5, 2.

b. Skin, unstuffed, bad state. Tail imperfect.
S. Africa. 52, 3, 24, 2.

Section 2. LORICATA.

The body covered with bony convolute armour, formed of bands or rings of *tesseræ*, generally revolute; teeth many.

Dr. Fitzinger, since he has left Vienna and gone to Munich, has occupied himself in compiling monographs of various groups of Mammalia, from different works, having no facility for examining any specimens; and a reader might be misled by the mention of the existence of certain species in the Museums of London, Leyden, Berlin, &c., but he only speaks of these specimens from books, from which the whole of his descriptions are taken, he never having visited any of these collections. He forms sundry new genera and species, from differences he thinks he has observed in the descriptions of different authors, and refers in several cases the same species to more than one genus. Several continental Zoologists have come to the determination of ignoring his monographs. There is no doubt that this course, if possible, would be for the advantage of Zoology, but unfortunately it cannot be followed. Some compiler will come, who, as is often the case, from ignorance of history, will consider all scientific papers, especially such as are published by a National Academy, as of authority, and quote them hereafter, and consider them as of equal authority with the most carefully prepared monographs founded on the examination of specimens.

Professor Gervais, in the 'Histoire Naturelle des Mammifères,' published in 1855, vol. ii. p. 252, describes and figures the change of teeth in *Tatusia peba;* and Professor Flower, in the P. Z. S., 1868, p. 380, confirms his observation by the examination of two young specimens in the Museum of the Royal College of Surgeons, and a larger skeleton in the collection of the British Museum, and concludes that the "animal almost attains the dimensions of the adult before the teeth are finally shed," and he complains that he is not "able to find a single specimen of the right age to throw any light on this question. All specimens are either too old or too young. With the exception of the one species above described, all statements with reference to the succession of the teeth of these animals appear to rest upon no sufficient basis of observation." The Museum specimen of the skull of *Praopus Kappleri* shows the same change.

Family 1. TATUSIADÆ.

Dasypodidæ, A, *Gray, Cat. Carniv. &c.* p. 377.

Dorsal disk closely attached to the back; divided in the middle by separate rings into a moderate scapular and pelvic shield. Head elongate, the ears close together on the top of the head. Toes of the fore and hind feet separate, covered with plates above. Claws conical. Tail with rings of shields. Skull separate from frontal plate. Nose elongate; intermaxillary bones triangular, shelving off from the upper part of the maxilla to the sides of the palate. Pelvic shield free from the pelvis.

1. TATUSIA, *Gray, Cat.* p. 377.

Face suddenly contracted, more subcylindrical, with two lines of *tesseræ* behind the eye, and more or less numerous shields on the sides of the cheek, before the eye. Under part of palate of skull rather narrow, flat, slightly shelving off, and rounded on the sides. Tail slender, all the rings formed of flat shields, the upper ones scarcely prominent in the middle of the hinder edge; the upper medial shields of the hinder bands like the others, or of the middle of the back of the last one, only rather shorter.

All the smooth-tailed and nearly bald *Tatusiæ* have been considered as one species, and those that come from different parts of the warmer parts of America are very much alike, and have been considered the same species. They differ in the form and comparative size of the head, and in the form of the *tesseræ* of the frontal shield, but the differences are so slight as scarcely to be defined by words. Their shields are also very much alike, but yet there are small differences, indicating that there are probably distinct species; but more specimens are wanted with more distinctly marked habits, to come to any certain conclusions on this subject. The most definite characters for the distinction of these species are, perhaps, afforded by the form of the lachrymal bone; that at least seems to furnish the most easily described character.

The lachrymal bone in some is nearly square, that is to say, broad and truncated in front; in others it is rounded in front; at other times it is more or less triangular, the upper edge being straight, and the front lower one gradually shelving to the lower part of the hinder side. The lower surface generally forms part of the lateral keel, on the zygomatic arch, but in *Tatusia peba* the whole of the keel is formed of the malar bone, the lachrymal **bone** only reaching to the upper part of the prominence.

Dr. Fitzinger, in the 'Sitzungsbericht,' **1871,** p. 335, describes as separate species, *Dasypus peba*, *D. urocerus*, *D. Lundii*, and *D. mexicanus;* and he forms a genus *Cryptophractus*, for *Praopus hirsutus* of Burmeister, but he only knows them from books.

In a very **young** specimen (113 *c*) in the British Museum **the head** is covered with naked skin, the shields not being developed. The dorsal zones are imperfectly developed, of about fifty-two *tesseræ*; six zones are complete, and the hinder ones are only developed on the side of the back. The rings of the tail are each formed of two series of shields, and far apart.

* *The lachrymal bone subtriangular. The lower edge just above the keel of the zygomatic process.*

1. Tatusia peba, *Gray, Cat.* p. 377.
Tatou noir, *Cuvier, Oss. Foss.* v. t. 10 (skeleton and skull, good).

Fourth of the free rings of the body with 55 *tesseræ*. Frontal shield rather rounded behind, with a transverse groove between the second and third series of *tesseræ*. Two narrow streaks of four or five shields below the eye behind. The sides of the face before the eyes with a few large, flat, broad *tesseræ*, in two elongated series, only slightly raised above the skin. Skull rather depressed, rather narrow. Forehead convex. Nose tapering, rounded, rather slender. Nasal bones moderate, narrow and contracted behind. Intermaxillaries shorter than half the distance from the end of the nose to the first grinder. Lachrymal bone large, subtriangular; upper edge nearly as long as the bone is high; front and lower edge curved to just above the blunt keel of the front lateral edge of the zygomatic process.

113 *b*. Animal, unstuffed, with 10 dorsal rings. Head and body 17
    in. Tail 11 in. Not in good state.
Skull of "*b*," wanting occiput.
    S. America.

113 *g*. Animal, stuffed, with 9 dorsal rings. Back with scattered
    hairs. Head and body 13 in. Head very narrow.
    Seba, i. t. xxix. f. 1 (moderately good).
Skull of "*g*."
    W. Indies. 42, 12, 27, 8.

113 *i*. Animal, stuffed, with 10 dorsal rings; slightly bristly. Head
    and body 10 in. Tail 8½ in.
Skull of "*i*," slightly developed. Teeth conical, not worn down.
    S. America.

911 *b*. Animal, stuffed. Head and body 11½ in. Tail 10 in.
Skull, showing succession of teeth, and skeleton.
    Tatusia, *Flower, P. Z. S.* 1868, p. 379.
    America. 46, 1, 19, 3. Warwick.

Animal, stuffed; young.   57 tesseræ in the dorsal rings.
America.   42, 12, 27, 7.   (Skull not seen).

Animal, stuffed; young.   60 tesseræ in the dorsal rings.
America.   42, 5, 10, 7.   (Skull not seen).

Animal, not stuffed; young.   Tail broken.
America.   45, 8, 31, 32.   (Skull not seen).

Animal, stuffed; young.   52 tesseræ in dorsal band.
America.   46, 8, 9, 23.   (Skull not seen).

Presented by the Haslar Hospital.

911 *f*. Skull, 3½ in., without lower jaw.
S. America.   56, 12, 10, 695.

** *Lachrymal bone square or subtriangular, the lower side forming
part of the front of the keel of the zygomatic process.*
† *Tesseræ on cheeks before the eyes small, numerous, convex.*

2. TATUSIA MEXICANA. .

The frontal shield of large *tesseræ*, the hinder part rather produced
and rounded; the two hinder series of *tesseræ* separated from the
others by a cross groove.   Sides of the head with two pairs of
lines of several oblong convex *tesseræ* behind the orbit, and with a
large patch of numerous small tubercles in three or four series before
the orbits.   Ears covered with large, close, granular tubercles,
with larger tubercles at the base of the front.   Lachrymal bone of
skull large, truncated in front, rather wider behind, and keeled
beneath.   Nose rather shorter than that of *Tatusia granadiana*, and
rather broader behind.   Nasal bones linear, very slightly broader in
front.   Brain-case scarcely contracted or marked with a groove
across the crown.   Orbital hole on the upper part of the zygomatic
cavity short and deep.

Head and body 17¼ in.   Tail 15½ in.

113 *k*. Animal, stuffed.   Body and head 16 in.   Tail 13½ in.
Skull of " 113 *k*."
Mexico.   43, 9, 11, 6.

The first free dorsal band narrowed on the sides, and with a band
about one-third of the circumference, interjected between the narrow
part of the first and the hinder part of the second band.   The three
or four bands next the front edge of the anterior portion of the dorsal
shield convex.   The interjected band is unlike on the two sides ; on
the right side it is continued, and of a considerable breadth to the
under margin of the disk ; on the left side it is short, and the first
band gets broader near the ventral edge of the disk.

†† *The tesseræ on the cheeks before the eye large, flat nearly level
with the skin, with a ridge of small scales projecting behind
under the eye.*

3. TATUSIA GRANADIANA.

Side of the head with two or three linear diverging lines of small,
flat, oblong, keeled shields.   The cheeks before the eye covered with

large, flat, imbedded shields in several lines, the medial line consisting of larger shields and forming a line under the orbits. Ears granular. The lachrymal bone broader than high, truncated in front. The lower side keeled. Nasal bone long, slender. Brain-case contracted and impressed with a groove across the crown at the upper part of the zygomatic cavity, which is short. Shields of the hinder part of the crown separated from those of the occiput by a transverse central groove. Crown shields in four pairs; the two hinder pairs larger and united by a straight central suture, the two other pairs diverging from a triangular, central crown-shield. Tail nearly as long as the body and head. The under side of the neck, body, and outside of legs with whitish hairs, arising from the front edge of sunken tubercles, which are largest on the under side of the body and legs.

a. Animal, stuffed. Head and body 20 in. Tail 18½ in.
Skull of "*a*," t. f.
Antioquia, Concordia. T. K. Salmon. 73, 3, 12, 2.

Prince Neuwied has a *Dasypus longicaudatus*, but I do not find it figured in the Museum copy of his *Abhandlungen*, and in the accounts of it that I have seen the tail is said to be shorter than the head and body. So that it cannot be this species. It is generally regarded as a name for *T. peba*.

4. TATUSIA LEPTORHYNCHUS.

Frontal shield very convex, broad, rather produced and acute behind between the ears. Side of the head behind the eyes with two concentric ridges, each formed of a single series of oblong convex *tesseræ*, those of the hinder series being the largest. The whole of the cheeks and under sides of the head covered with large imbedded flat shields, those immediately behind and under the orbit being elongate and convex. Ears covered with scales. Skull depressed behind. Forehead convex, strongly constricted behind. Nose elongate, slender. Nasal bones very slender, attenuated behind; rather more than half the distance from the end of the nose to the front lateral tooth. Lachrymal bone large, nearly square, nearly as high as long on the upper edge; front edge rather arched backwards, the lower one gradually tending backwards on the keel. Temporal fossæ small, short, higher than long.

a. Animal, stuffed. Body and head 18½ in. Tail 16½ in.
Skull of "*a*."
Guatemala. Salvin. 65, 5, 18, 42.

5. TATUSIA BREVIROSTRIS.

Frontal shield very convex and rounded, rather produced behind, and with a distinct cross groove. The side of the head behind the eyes with two distinct ridges of *tesseræ*, the upper one longest, and formed of one series of convex *tesseræ*, the hinder one of two series of small ones. The cheeks on the front of the eye with large, naked, unequal, flat tesseræ in several rows. Under side of the head covered with large imbedded plates in several series. Skull rather depressed, behind very broad. Forehead convex, strongly constricted behind. Nose short, depressed, rather broad. Nasal bones broad, as wide before as behind; intermaxillaries scarcely half as long as the distance from the end of the nose to the first grinder. Lachrymal bone large, subtriangular, nearly as high as long, on the upper edge; front and lower edge arched, gradually contracted towards the hinder part on or just below the lateral keel. Temporal fossæ moderate, short, higher than long.

a. Animal, stuffed.
911 e. Skull of "a," t. f. 46, 5, 13, 16.
    Rio Janeiro. Clausen. 44, 3, 7, 2.

911 c. Skull, 3¾ in.
    Bolivia. Bridges. 46, 5, 13, 16.

6. TATUSIA LEPTOCEPHALA.

Head-shield and scutella of side of head unknown. Skull long and slender, gradually tapering in front. Nasal slender, elongate. Intermaxillaries rather less than half the length of the distance from the end of the nose to the front tooth. Forehead convex, rounded. Lachrymal bone small, triangular above in front; very strongly keeled on the side, and concave below. Temporal fossæ small, subcircular; incomplete behind.

911 c. Skeleton without dorsal or head shields.
    Brazils. 47, 4, 6, 2.

7. TATUSIA BOLIVIENSIS.

Animal unknown.
    Lachrymal bone truncated and slightly rounded in front, the lower side keeled. The nasal bones linear, very slightly broader in front. The brain-case contracted and impressed with a groove across the crown, at the upper part of the zygomatic cavity, which is short.

911 d. Skull.
    Bolivia. Bridges.

911 a. Skull and skeleton; very like "911 d" in general form, but the impressed line across the forehead is not so distinct, the zygomatic arch is longer, and the suture between the squamal and frontal bone is more in advance.

2. PRAOPUS, *Gray, Cat.* p. 379.

Face attenuated. Nose elongate-conical. Hinder part of palate of skull broad, concave, with raised edges on the side. The tail thick at the base, attenuated, the shields on the upper part, especially of the four or five basal rings, convex, and prominent on the hinder edge, the upper medial shields of the three hinder bands, especially of the hindermost one, much shorter and smaller than the lateral shields.

1. PRAOPUS KAPPLERI.

Tatusia Kappleri, *Gray, Cat. Carniv. &c.* p. 379.
Dasypus Kappleri, *Krauss, Arch. für Naturg.* 1862, p. 24; *Fitz. l.c.,* p. 356, t. iii. figs. 1 & 2 (skull).

1493 *a*. Animal, stuffed; adult. Head and body, 12 in. Tail, 16 in.
Dasypus Kappleri, *Krauss, Arch. Naturg.* 1862, p. 24, t. 3.
1493 *a*. Skull, 5 in., of "*a*."
    Surinam. Krauss.

1493 *b*. Animal, stuffed.
    Surinam. Krauss. 73, 4.

Skull. Nose tapering from the front of the zygomatic arch. Nasal
bones elongate, rather narrower behind, half as long as the distance
from the top to the front of the zygomatic arch. Intermaxillary bones
broad in front, gradually narrowed behind; longer than half the distance from the tip to the front of the zygomatic arch. Lachrymal
bone subtriangular, upper side elongate, produced forwards, lower side
gradually contracting for half its length, and then suddenly becoming
broader, with a rounded front edge, nearly as broad as the hinder end.
Palate broad, rather concave, with a sharp raised edge on each side
from just behind the hinder grinders.

Family 2. DASYPODIDÆ.

Dasypodidæ B, *Gray, Cat. Carniv. &c.* p. 376 & 379.

Dorsal disk attached by the skin to the sides of the back of the
    animal, and divided in the middle by separate rings into a
    moderate scapular and pelvic shield. Head moderate, more
    or less broad. Ears far apart, lateral. Toes of front feet
    united to the claws, compressed. Claws compressed, sharp-
    edged below, bent inwards. Toes of hind feet separate. Skull
    conical; intermaxillary bone quadrangular on the sides, trun-
    cated behind. Pelvic shields free from the pelvis. The first
    and second finger slender, weak; the third and fourth shorter,
    thicker; the fifth similar to the second, but shorter or rudi-
    mentary. Walks with the fore feet flat on the ground.

Tribe 1. DASYPODINA.

Dorsal shields with six or eight free rings in the middle, and with
    two or three bands consisting of a few shields between the
    head and front of the frontal shield. Shields of tail placed in
    rings, and well developed. Skull depressed, spread out, very
    broad behind, conical in front. The digital bones of the first,
    second, and third fingers moderately elongate; those of the
    third the thickest but somewhat shorter than the bones of the
    fourth and fifth fingers. Middle ~~hind~~ claw very large.— *front*
    Flower, Manual Osteol. p. 276, f. 101.

    * *Intermaxillaries with a tooth on each side behind.*

1. DASYPUS.

Intermaxillaries with a tooth on each side behind. Frontal shield
of many *tesseræ*. The dorsal shields with two short bristles at the
hinder edge of each *tessera*; a cross row of large *tesseræ* behind. Under
side of body with scattered bristles.

1. DASYPUS SEXCINCTUS, *Gray, Cat. Carniv. &c.* p. 381.
Encoubert, *Cuv. Oss. Foss.* v. t. xi. f. 4 & 6 (skull).
D. sexcinctus, *Rapp, Edentata*, t. iii. f. 4 & 5 (skull).
Euphractus setosus, *Fitzinger, Sitzungsberichte*, p. 251.

Var. 1. *Nose broad.* **Front** *dorsal shield with only 5 bands in the centre of the back.*

114 *a*. Animal, stuffed; adult. Not in good state. Body and head, 14 in.
S. America.   Presented by Lieut. Mawe.

114 *b*. Animal, stuffed; adult. Large.
Dasypus encoubert, *Zool. Soc.*
S. America. 58, 9, 4, 12.

114 *c*. Animal, stuffed; adult. Body and head, **16 in.**
Skull of " *c*." 48, 19, 5, 1.
S. America. 43, 9, **15**, 7.

Var. 2. *Nose convex, narrow. Front dorsal shield with 4 bands in the middle of the back.*

729 *h*. Skin, unstuffed. Head and body, 18 in.
S. America, Bolivia. Bridges. 46, 7, 28, 11.

729 *d*. Animal, stuffed. Head and body, 15 in.
S. America. 43, 12, 6, 15.

729 *g*. Animal, stuffed. Head and body, 17 in.
b. Skull of " *g*." End truncated. 46, 6, 15, 3.
S. America. 46, 4, 25, 26.

729 *e*. Animal, stuffed. Head and body, 13 in.
S. America. 43, 12, 6, 16.

729 *f*. Animal, stuffed. Head and body, 11 in.
S. America. 46, 12, 13, 20.

729 *c*. Skeleton of adult.
Dasypus sexcinctus.
S. America.

729 *d*. Skeleton.
" D. octodecimcinctus."
S. America. Zool. Soc. **51, 8, 16, 24.**

729 *a*. Skull.
S. America. 43, 10, 5, 1.

729 *i*. Skull covered with head shields.
S. America. 55, 3, 11, 6.

2. CHÆTOPHRACTUS.

Intermaxillaries **with a tooth on each** side behind. Dorsal shield with several long hairs at **the hinder edge** of each *tessera*; under part of body very hairy.

1. CHÆTOPHRACTUS VILLOSUS.

Euphractus villosus, *Gray, Cat. Carniv. &c.* p. 382.
Chætophractus villosus, *Fitz. Sitzungsb. Akad. d. Wissen. Wien.* p. 268.
D. villosus, **Giebel, Zeitsch. 1861**, f. 1 in t. iii. iv. v. (skull).

914 *b*. Animal, stuffed. Head and body, 16 in.
b. Skull of "*c*."
S. America, Pampas.

914 *a*. Animal, stuffed. Head and body, 10 in.
"Quirquindis peludo," Bridges.
S. America. Bolivia. Bridges. 47, 11, 22, 18.

914 *c*. Skin of back. Head and body, 14 in.
Buenos Ayres. Zool. Soc. 47, 5, 17, 11.

914 *a*. Skeleton, unmounted. Skull punctulated.
S. America. 48, 11, 18, 5.

In the *P. Z. S.*, 1865, it was thought that Giebel had figured the
skull of *E. sexcinctus* under the name of *E. villosus*, and therefore
D. villosus was referred to *Euphractus*, but now, having obtained the
skull of this species, it appears that Giebel was right, and that the
skull differed from *D. sexcinctus* in having a much more slender nose.
Fitzinger, in his compilation, refers *D. sexcinctus, vellerosus* and
villosus to different genera, erroneously saying that the two former
are provided with teeth on the intermaxillaries, and that *D. villosus*
and *D. minutus* belong to another genus.

2. CHÆTOPHRACTUS VELLEROSUS.
Dasypus vellerosus, *Gray, Cat. Carniv. &c.* p. 361.
Dasyphractus brevirostris, *Fitz. Sitzungsberichte*, p. 264.
Cryptophractus brevirostris, *Fitz. Voy. Novara*, p. 295.

457 *a*. Animal, stuffed.
Dasypus vellerosus, *Gray, P. Z. S.* 1865, p. 376, t. 18.
457 *a*. Skull of "*a*."
Santa Cruz de la Sierra. Bridges.

*** Intermaxillaries without any tooth on the hinder edge.*

3. EUPHRACTUS.
Intermaxillaries without any tooth on the hinder edge.

1. EUPHRACTUS MINUTUS, *Gray, Cat. Carniv. &c.* p. 382, f. 43
(skull).
Chætophractus minutus, *Fitzinger, Sitzungsberichte*, p. 272.

139 *a*. Animal, stuffed. Head and body, 11½ in.
S. America.   Presented by Capt. Fitzroy.

139 *b*. Animal, stuffed. Ears wanting. Head and body, 13 in.
S. America. 31, 12, 29, 22.

139 *c*. Animal, stuffed. Head and body, 10¾ in.
S. America.   Presented by W. Reeves, Esq.

139 *d*. Animal, stuffed. Head and body, 11 in.
Buenos Ayres. 55, 12, 24, 288. Presented by Charles Darwin, Esq.

139 *e*. Animal, stuffed. Head and body, 10¾ in.
912 *a*. Skull of "*e*."
Gray, P. Z. S. 1865, p. 377, fig.; *Cat.* p. 383, f. 45.
Bolivia. Bridges. 45, 11, 18, 34.

912 *f.* Animal, stuffed. Head and body, 9½ in.
S. America. Zool. Soc. 55, 12, 24, 287.

912 *g.* Animal, stuffed. Head and body, 8¾ in.
S. America, Bolivia. Bridges. 45, 11, 18, 33.

912 *h.* Animal, stuffed. Head and body, 10¼ in.
S. America.

912 *i.* Animal, stuffed. Head and body, 11¾ in.
S. America.   Presented by the Haslar Hospital.

912 *b.* Skeleton, and dorsal and frontal shield, 11½ in., and tail over it.
S. America. Zool. Soc. 67, 7, 8, 8.

Tribe 2. PRIONODONTINA.

Central dorsal rings numerous, 12 or 13. Shields of tail placed
alternately, often rudimentary. The third and fourth outer
hinder claws very large. Skull plano-convex, high, swollen;
front constricted in the middle. Nose subcylindrical, tapering.
Intermaxillaries elongate, truncated behind; toothless. The
digital bones of the first and second fingers very slender, of
the third and fourth very thick; the lower ones very short.
The last bone, especially of the third finger, very long compared
with the others. The fifth finger rudimentary, reduced to a
single bone.—*Flower, Manual Osteol.* p. 276, fig. 102.

    * *Teeth many, small. Tail with rings of well-developed shields.
Intermaxillaries broad, square behind.*

4. PRIONODOS.

Central free rings 13; anterior shield of 10—11, and hinder of 16
rings.

    1. PRIONODOS GIGAS, *Gray, Cat. Carniv. &c.* p. 380.

Cheloniscus gigas, *Fitz. Sitzungsberichte*, 1871, p. 227.
D. gigas, *Cuvier, Oss. Foss.* v. p. 128, t. xi. f. 1—5 (skeleton).
Rapp, Edentata, t, iv. *b.* (skeleton).

76 *a.* Animal, stuffed. Head and body, 34 in.
S. America. Bullock's Museum.

76 *b.* Animal, stuffed, 37 in.
Chili.   Presented by Lieut. Mawe, R.N.

76 *a**. Skeleton, mounted.
S. America.

76 *b.* Skeleton, unmounted.
S. America.

76 *c.* Skull, 7¼ in.
Brazils. Brandt. 46, 4, 21, 6.

^{**} *Teeth moderate, 9 or 10 on each side. Tail naked, with sunken tubercles. Intermaxillaries obliquely truncated behind, the lower margin longer than the upper.*

5. XENURUS.

Head covered with many (50—52) bony shields, with a single odd shield behind. Ears covered with scales. The central dorsal rings, 12 or 13, consisting of 30—31 shields. Anterior shield of 7 and hinder of 9 rings of shields.

1. XENURUS UNICINCTUS, *Gray, Cat. Carniv. &c.* p. 384.
Cabassou, *Cuvier, Oss. Foss.* vol, v. p. 120, t. xi. f. 7—9 (skull).
X. verrucosus, *Fitz. Sitzungsb. Akad. d. Wissen. Wien.* 1871, p. 233.
X. loricatus, *Fitz. Sitzungsberichte,* p. 239 ?
X. gymnurus, *Fitz. Sitzungsberichte,* p. 242.

The head covered with a number (50—52) of shields, arranged in a cross series on the occipital edge, and with one odd shield in front of them, surrounded by six shields, and with five pairs of shields in front of them. Two concentric rows of plates round the central ones, those of the inner series being the largest. Cheeks in front of the eye with three rows of small close shields. Three free rings of shields before the front of the dorsal disk, each formed of equal-sized, well-developed shields. The front zone consisting of five shields, the middle one the largest and the outer the smallest. The second of six nearly equal-sized square shields, with one or two small round shields on the outside of them. The third formed of 10 shields, the four central being large and square, and the six of the outer ends smaller, the outermost being very small. The movable zones of the back composed of 30—31 shields, which often have a thickened margin on each side. The hinder part of the dorsal shield ending in a semicircular notch; the skin which covers this part and the base of the tail studded with small, roundish or oblong, distant, immersed, hard tubercles. The ears large, covered externally with close scales, which are larger at the edge. The whole back has a quantity of short hairs projecting between the shields, and the under side of the head and body is covered with long black hairs, which are most abundant on the throat.

731 *b.* Animal, stuffed. Head and body, 23 in.
b. Skull of "*b,*" 4½ in. ; imperfect. 46, 4, 25, 6.
S. America, Brazils. Clausen. 44, 3, 7, 1.

731 *c.* Animal, stuffed. Head and body, 20½ in.
Skull of "*c,*" 3½ in. End imperfect. 46, 6, 15, 12.
D. gymnura, ♀, *Brandt.*
S. America, Brazils. Brandt. 46, 1, 9, 23.

731 *d.* Animal, stuffed. Head and body, 18 in.
Dasypus gymnura, ♂, *Brandt.*
S. America, Brazils. Brandt. 46, 1, 9, 24.

731 *e.* Animal, stuffed; male. Head and body, 17¼ in.
731 *e.* Skull of "*c.*" End imperfect. 46, 5, 13, 15.
Brazils. 45, 8, 25, 16.　　　　　Presented by R. Graham, Esq.

731 *f*. Skin, not stuffed.   Head and body, 23½ in.   Not in good state.
S. America.

731 *g*. Animal, stuffed.   Head and body, 20½ in.
Brazils, St. Catherine's.   Parzudaki.   51, 8, 25, 11.

731 *d*. Skeleton, unmounted.
D. gymnurus.
S. America, Brazils.   Becker.   47, 4, 68.

731 *f*. Skeleton, unmounted.   Skull, 4 in.
Dasypus gymnurus.
S. America.   Brandt.   48, 5, 6, 15.

731 *e*. Skeleton, unmounted.
Dasypus gymnurus.
S. America.   Brandt.   47, 4, 5, 6.

731 *a*. Skull, behind imperfect, 4 in.
S. America.

2. XENURUS LATIROSTRIS.

Dorsal shields with 30 rings; the *tesseræ* square, with raised edges,
which are only on the lateral edges of the short broad *tesseræ*, of
which the movable rings are composed.   A series of short white hairs
coming out from the back edge of the *tesseræ*.   The under side of the
body and outside of the legs and thighs with longer, more abundant,
hair.   The outside of the legs and thighs with distant large scales.
Feet covered with closer oval scales.   Tail with imbedded small
plates.   The nose of the skull is very broad, short and swollen at the
side.   The hinder part of the zygomatic arch is very broad, much
dilated, and united by a long, straight suture to the prominence of the
os petrosum.   The articular process of the lower jaw short, nearly
erect, and separated from the front upper process by a semicircular
notch.   The nose short and broad, with a convexity on each side.

1597 *a*. Animal, stuffed; male.   Head and body, 15½ in.
Skull of "*a*."
Xenurus unicinctus, *var.*, *Gray, Cat.* p. 314.
Brazils, St. Catherine's.   51, 8, 25, 9.

6. ZIPHILA.

The head covered with 30—32 large shields, with three odd shields
in an interrupted central line.   Ears nakedish, with few scales on the
margin.   The central rings of the body (12—13) consisting of 27
shields.   Head elongate; crown convex, with a few large shields.
Four odd shields in the central line.   The hinder very large, transverse; the second small, oblong, transverse, four-sided; the third hex-
agonal, longer than broad.   The second hinder central crown shields
surrounded by 3 pairs of large shields, the front pair being the longest;
all surrounded by another series of 6 large shields, which are themselves surrounded by a series of much smaller shields.   The cheeks in
front of the eye with a few scattered, distant, round shields.   The ears
nakedish, rather scaly on the edge.   The three rings in front of the

back free, but scarcely divided from the first rings of the back itself; the front one formed of seven small transverse shields, the central one circular, and the lateral ones oblong. The second ring formed of three or four oblong shields on each side, the central pair being the largest, and the two outer smaller. The third band formed of six shields, the four middle ones of which are largest, and like the shields of the back. The front and hinder shields of the dorsal disk large and square; of the rings, oblong. The back with 27 plates in the cross-band. Tail conical, nakedish, with scattered, round, flat, immersed shields. The front foot with two slender internal toes and claws, and three broad strong external claws, the middle one much the longest. Skull tapering in front, like *Xenurus*. Upper grinders, 6—6, the two front ones on each side probably wanting, but there are no signs of them; lower grinders, 8—8, as in *Xenurus*. The hinder part of the zygomatic arch not nearly so broad as in all the specimens of *Xenurus*, and united to the process of the *os petrosum* by a much shorter suture. The sides of the nose only slightly convex. The articulated processes of the lower jaw bent backwards, and separated from the front process by a very wide space, which has an arched notch at the front end. The skull, which appears much older than any larger skulls of *Xenurus tricinctus*, is only 3⅗ in. long, and 1⅗ in. broad. The Kabassou of Buffon, Hist. Nat., vol. x. t. 40, appears from the large size of the scutella on the forehead to belong to this genus, but the tail is more that of a *Dasypus*.

1. ZIPHILA LUGUBRIS.

1598 *a*. Animal, male. Head and body, 16 in. Badly preserved. Skull of " *b*."

 Brazils, St. Catherine's. 51, 8, 25, 10.

1598 *b*. Animal, stuffed. Head and body, 12½ in.

 Xenurus unicinctus, *var.*, *Gray, B. M.*

 S. America, Demerara. Snellgrove. 55, 8, 28, 9.

The two specimens are in very different conditions, one (*a*) having very hard stony shields, and the other (*b*), which has, perhaps, been kept in confinement, very thin weak shields; but in other respects they agree in most of their characters. The one with thin shields has the two front bands of shields before the dorsal disk scarcely deve-loped, and its head-shields are rather irregular compared with the other specimen. There is a small, square, odd, central shield between the occipital and crown-shield of the other specimen.

Professor Burmeister refers to the Tatu, Seba Thes. i. t. 30, f. 3 & 4, for this species as representing *Dasypus hispidus*. It certainly agrees with this in having fewer shields on its head than *X. unicinctus*, but neither head (and those of the two figures are very different from one another) is like our specimens.

<h2 style="text-align:center">Family 3. TOLYPEUTIDÆ, Gray, Cat. Carniv. &c. p. 385.
Tolypeutes, Gray, Cat. Carniv. &c. p. 385.</h2>

1. CHELONISCUS.

Head-shield broad, rounded, not produced behind, with three shields in middle of hinder part.

1. CHELONISCUS TRICINCTUS.
Tolypeutes tricinctus, *Gray, Cat. Carniv. &c.* p. 386, f. 46 (skull).
Fitz., Sitzungsb. Akad. d. Wissen. Wien. 1871, p. 869.

730 *a.* Animal, stuffed. Head and body, 14 in.
b. Skull. 49, 3, 13, 1.
     Gray, P. Z. S. 1865, p. 379, f. ; *Cat.* p. 386, f. 4—6.
     S. America. 49, 3, 12, 1.

2. TOLYPEUTES.

Head-shield narrow, produced behind, with two shields, one behind
the other in the centre.

The dorsal shield of *T. conurus* is quite free from the body of the
animal, except in three places, where it is united to it by the skin of
the animal, which from these three places extends over the whole of
the inner surface of the disk ; 1st, over the back of the neck, which
is united over the front end of the scapular disk ; 2nd, on the sides of
the body to the outer ends of the three medial dorsal rings ; 3rd, at
the pelvis, which is united to the caudal end of the pelvic disk,
leaving the tail free. (See Ann. & Mag. of Nat. Hist. 1873, May).

1. TOLYPEUTES CONURUS, *Gray, Cat. Carniv. &c.* p. 386.
Dasypus conurus, *Giebel, Zeitsch.* 1861, f. 2, of t. iii. iv. v. (skull,
lower jaw much broader than in our specimen).
Sphærocormus conurus, *Fitz. l. c.* p. 376.

140 *a.* Animal, stuffed ; adult.
     Tatusia tricincta, *Gray, Cat. Mamm.* 1843, p. 189.
     S. America. 1837.   Presented by Capt. Fitzroy.

140 *e.* Animal, stuffed. Head and body, 14½ in.
     S. America.

140 *f.* Animal, not stuffed ; male, adult.
     S. America.

140 *g.* Animal, not stuffed ; male, adult.
     S. America.

140 *h.* Animal, stuffed ; female. Head and body, 12 in.
Skull of " *h.*"
     S. America.

140 *b.* Dorsal shield, head, and tail.
     T. tricincta, *Gray. Cat. Mam.* 1843.
     S. America.

140 *c.* Dorsal shield, head, and tail, 14¼ in.
     T. tricincta, *Gray, Cat. Mam.* 1843.
     S. America. 41, 602.

140 *d.* Skull, frontal and dorsal shield.
     S. America.

Family 4. CHLAMYDOPHORIDÆ, *Gray, Cat. Carniv. &c.* p. 387.

The dorsal disk divided into two parts behind, consisting of an elongated dorsal and a short pelvic shield, the latter attached to the bones of the pelvis.

1. CHLAMYDOPHORUS, *Gray, Cat. Carniv. &c.* p. 388.

The dorsal disk only attached by the central line to the middle of the back, which is covered with hair.

1. CHLAMYDOPHORUS TRUNCATUS, *Gray, Cat. Carniv. &c.* p. 388. *Fitz. l. c.* p. 382.

1048 *a.* Animal, stuffed.
a. Skeleton of "*a.*"
  C. truncatus, *Yarrell, Zool. Journ.* iii. p. 544, t.
  Mendoza.   From the Mus. Zool. Soc.

1048 *c.* Animal, in spirits.
  Mendoza.   71, 5, 19, 11.

1048 *b.* Skeleton.
  C. truncatus, *Gray, P. Z. S.* xxv. 1857, p. 9 (skeleton).
  Mendoza.

Section 3. VERMILINGUA.

Body covered with hairs or spines.   Head elongate, mouth small, tongue elongate.   Teeth none or numerous.

Family 1. ORYCTEROPODIDÆ, *Gray, Cat. Carniv. &c.* p. 389. Teeth numerous.

1. ORYCTEROPUS, *Gray, Cat. Carniv. &c.* p. 389.

Covered with scattered bristly hair ; longer on the haunches.

1. ORYCTEROPUS CAPENSIS, *Gray, Cat. Carniv. &c.* p. 389. *Cuv. Oss. Foss.* v. p. 117, t. xii. (skeleton and skull).   *Rapp, Edentata,* p. 13, t. i.—iv. (skull) ; *Duvernoy, A. S. N.* xix. p. 192, t. 96 (skull).   *Sclater, List of Vertebrated Animals,* 1872, p. 112, f. 16.

72 *a.* Animal, stuffed, 5 ft. 9 in.
  Cape of Good Hope.   Presented by W. Burchell.

72 *b.* Animal, stuffed, 5 ft. 9 in.
  Cape of Good Hope.

72 *c.* Animal, stuffed ; female.   Young, 26 in.
  Cape of Good Hope.   Verreaux.   1837.

72 *e.* Skin of adult, unstuffed.
931 *a.* Skull of "*e.*"
  Cape of Good Hope.   43, 3, 25, 8.

931 b. Skeleton.
  S. Africa.

2. Orycteropus senegalensis, *Gray, Cat. Carniv. &c.* p. 389.
Duvernoy, Ann. Sci. Nat. xix. p. 122, t. ix.

72 *d.* Animal, stuffed; bad state; short ears.
D. capensis, "*d*," *Gray, Cat.* p. 190.
W. Africa. 39, 12, 26, 1.

72 *f.* Skull and bones of feet, from bad skin.
W. Africa. Randall.

3. Orycteropus æthiopicus, *Gray, Cat. Carniv. &c.* p. 389.
Sclater, P. Z. S. 1872, p. 669, fig. *List V. Animals,* p. 113, f. 17.
Sundevall, V. Akad. Handl. 1841, p. 226, t. 3, f. 1—5 (animal and skull).

a. Animal, stuffed, 5 ft. 8 in.
E. Africa.

These species differ very much in the state of the fur. In *O. capensis* it is covered with black hair, especially in the hinder part of the back. *O. senegalensis* is dark-coloured, with long fur on the outside of the thighs, and the one from E. Africa is paler and covered with shorter hair.

Family .2 MYRMECOPHAGIDÆ, *Gray, Cat. Carniv. &c.* p. 390.

Skull with an incomplete zygomatic arch, the malar elongate, only articulated with the maxilla in front, and not reaching the very short zygomatic process of the squamosal.

Tribe 1. MYRMECOPHAGINA.

Terrestrial. Tail bristly. Walks on the side of the feet, with claws curved up. Malar bone slender, styliform, *Flower, Man. Ost.* p. 204, f. 65.

1. MYRMECOPHAGA, *Gray, Cat. Carniv. &c.* p. 390.

1. Myrmecophaga jubata, *Gray, Cat. Carniv. &c.* p. 390.
Flower, Man. Osteol. p. 204, f. 65 (skull).
Tamanoir, *Cuv. Oss. Foss.* vol. v. t. ix.

62 *d.* Animal, stuffed, 7 ft. long.
S. America.

62 *b.* Animal, stuffed, 43 in.
Columbia. Parzudaki. 45, 9, 18, 8.

62 *e.* Animal, stuffed, 38 in.
S. America. 49, 4, 26, 1.

62 *c.* Animal, stuffed; young.
Guiana. Stuffed and presented by C. Waterton, Esq.

1072 *a.* Skeleton.
S. America.

1072 *b.* Skull.
La Plata, Bravard.

1072 *c.* Skull, imperfect; from "62 a," *Cat. Mam.*
S. America. From Bullock's Museum.

Tribe 2. TAMANDUINA.

Skull elongate, slender; nose very long, slender. Brain-case elongate, narrow. Neck elongate, formed of elongate vertebræ. Ribs strong, rather dilated on the outer side behind. Tail conical, prehensile. Toes 4—4; the two middle front ones large.

2. TAMANDUA, *Gray, Cat. Carniv. &c.* p. 391.

1. TAMANDUA BIVITTATA, *Gray, Cat. Carniv. &c.* p. 391.
M. tamandua, *Blainv. Osteogr.* t. (skeleton, ¼).

Var. 1. Opisthomelas. *The sides and hinder part of back to the base of the tail black.*

40 *a*. Animal, stuffed. Dorsal streak broad to loins.
Brazils.   Presented by Lieut. Mawe, R.N.

40 *b*. Animal, stuffed. Dorsal streak narrow to the middle of the back.
Brazils.

40 *c*. Animal, stuffed; half-grown. Back grey-washed.
733 *c*. Skeleton of "*c*."
S. America. Zool. Soc.

40 *d*. Animal, stuffed; young. Back greyish.
Brazils. Zool. Soc. 54, 12, 6, 2.

40 *e*. Animal, unstuffed.
Brazils, Orinoco. Parzudaki. 44, 5, 13, 45.

40 *r*. Animal, stuffed. Back grizzled.
Brazils.   Presented by the Haslar Hospital.

40 *q*. Animal, unstuffed; young. Colour rather diffused.
Brazils. 44, 5, 14, 30.

Var. 2. *Brown or blackish, the parts that are white in the other varieties being only rather paler or indistinctly seen.*

40 *m*. Animal, not stuffed. Pale brown. Back rather darker, grizzled.
S. America. 50, 5, 24, 1.

40 *n*. Animal, stuffed; yellowish. Back blackish, brown-grizzled.
Brazils. 45, 9, 18, 9.

40 *o*. Animal, not stuffed; adult. Blackish-grizzled. Head, neck, and legs paler.
Brazils.

40 *p*. Animal, not stuffed; young. Blackish-grizzled on the head, neck, and limbs.
Brazils.

Var. 3. Opistholeuca. *Rump to the middle of the back white.*

Tamandua tetradactyla, *Sclater, P. Z. S.* 1871, p. 246, t. xliii.
40 *f*. Animal, unstuffed; adult. Streak of back continued to the white rump. Head and body, 26 in. Tail, 22 in.
New Grenada. 50, 7, 8, 39.

40 g. Animal, unstuffed ; adult.
Nicoya, Costa Rica. Salvin. 65, 5, 18, 12.

40 h. Animal, unstuffed; adult. Tail imperfect.
Guatemala, Duénas. Salvin. 65, 5, 18, 11.

40 i. Animal, unstuffed. Black, greyish-washed.
Guatemala. Salvin. 68, 7, 9, 2.

40 j. Animal, unstuffed ; young. Greyish-washed.
Vera Paz, Guatemala. Salvin. 65, 5, 18, 13.

733 a. Skeleton. Skull, 5¼ in. Without zygomatic arch and intermaxillary bone.
Tropical America. Brandt.

733 b. Skull, 5⅛ in. With intermaxillaries and zygomatic arch.
Amazons? Bates. 53, 3, 19, 74.

40 c. Young, in spirits.
Brazils.

40 k. Animal, unstuffed. Back of head and neck like back, reddish-
brown.
"M. annulata," Verreaux ; not Desmarest.
New Granada. Verreaux. 54, 9, 20 3.

40 l. Animal, young. Tail injured at the end.
S. America. 56, 1, 10, 1. "Mataperro or Dog-Killer."

2. TAMANDUA LONGICAUDATA, Gray, Cat. Carniv. &c. p. 392.

a. Animal, stuffed ; female. Body and head, 36 in. Tail, 18 in. = 54 in.
Skull of "a."
S. America. 46, 6, 1, 34. Parzudaki.

b. Animal, unstuffed ; male. Body and head, 24 in. Tail, 24 in.
= 48 in. Bones of one leg extracted.
Tropical America. 44, 1, 18, 10.

Tribe 3. CYCLOTHURINA.

Arboreal. Fingers two, of but two phalanges. Second finger
rather slender; third very thick, with a very strong claw ; first,
fourth and fifth abortive, the fourth rudimentary, united to the
metacarpal of the second. Toes 5, slender, subequal. Tail
prehensile. Ribs 16, the latter eight floating, all with a broad
hinder margin, so as to imbricate one over the other. Skull, face
short conical ; brain-case very large, swollen. Neck short,
formed of very short vertebræ. Tail conical, naked beneath,
prehensile.

3. CYCLOTHURUS, Gray, Cat. Carniv. &c. p. 392.
  1. CYCLOTHURUS DIDACTYLUS, Gray, Cat. Carniv. &c. p. 392.
  M. didactyla, Rapp, Edentata, p. 15, t. v.; Blainville, Osteol. t.
  (skeleton, n. s.)

3 c. Animal, stuffed ; not good state.
S. America.

5 *d.* Animal, stuffed.
S. America. 37, 4, 28, 44.

5 *e.* Animal, stuffed; bleached.
S. America. 38, 12, 29, 21.

5 *a.* Animal, in spirits.
Brazils.

5 *b.* Animal, in spirits.
British Guiana.   Presented by Dr. Hancock.

5 *f.* Animal, in spirits.
British Guiana. 62, 12, 15, 106.

3 *f.* Animal, unstuffed. Fur brown, varied above, with a central
blackish ventral streak.

734 *b.* Skull of *f.*
S. America.

734 *c.* Skeleton, mounted.
S. America.

2. CYCLOTHURUS DORSALIS, *Gray, Cat. Carniv. &c.* p. 392.

a. Animal, not stuffed. Tail imperfect.
Guatemala. Vera Paz. Salvin. 48, 2, 10, 1.

c. Animal, stuffed, young. Mark on back, sides and head brown.
S. America. Jeude. 58, 4, 28, 14.

b. Skin, not stuffed, without legs.
Costa Rica. Salvin. 65, 5, 18, 14.

d. Animal, in spirits.
Costa Rica. Zool. Soc. 55, 12, 26, 345.

e. Animal, in spirits, young.
Costa Rica.

Section 4. MONOTREMATA, *Gray, Cat. Carniv. &c.* p. 393.

Animal covered with hair or spines. Mouth small, tongue elongate.
Pelvis with marsupial bones; and a merrythought like birds. Hind
feet of male spurred.

Family 1. ORNITHORHYNCHIDÆ, *Gray, Cat. Carniv. &c.* p. 393.

1. PLATYPUS, *Gray, Cat. Carniv. &c.* p. 393.

1. PLATYPUS ANATINUS, *Gray, Cat. Carniv. &c.* p. 393.
Ornithorhynque, *Cuv. Oss. Foss.* v. t. xiv.

110 *f.* Animal, stuffed, 24 in.; male.
Van Diemens' Land. Gould. 41—1163.

110 *o.* Animal, stuffed, 25 in.; male.
S. Australia. 43, 3, 12, 54.   Presented by Capt. G. Grey.

110 *p.* Animal, stuffed, 23 in.; male.
Australia.   Presented by the Haslar Hospital.

110 *a*. Animal, stuffed, 23 in.; male.
Australia. Presented by Miss Reeves.

110 *h*. Animal, in spirits; adult.
Australia.

110 *i*. Animal, in spirits. Cut open and examined by De Blainville.
Australia.

110 *d*. Animal, stuffed, 18½ in.; male.
Australia. Presented by Dr. Merriman.

110 *b*. Animal, stuffed, 16¼ in.; female.
Australia.

110 *e*. Animal, stuffed, 15 in.; male. Discoloured from having been
in spirits.
O. rufus, *Leach, Zool. Miscel.* p. 136.
Australia.

110 *c*. Animal, stuffed, 14 in.; female. Sent as the specimen described
by Dr. Shaw, but Shaw's specimen was a male with spurs, and
this is a female without spurs.
Australia. Presented by R. Latham, Esq.

110 *g*. Animal stuffed, 13¾ in.; male.
Australia. Cuming. 56, 12, 3, 4.

110 *l*. Animal, unstuffed, 25 in.; male.
Australia. 52, 3, 29, 1. Presented by R. Howe, Esq.

110 *m*. Animal, unstuffed, 24 in.; female.
Australia. 47, 11, 13, 2. Presented by G. Duncan, Esq.

110 *k*. Animal, stuffed, 23½ in.; female.
Australia. Mantell. 43, 7, 1, 1.

110 *n*. Animal, unstuffed, 23 in.; female.
Australia. 44, 9, 30, 23. Presented by the Hon. E. India Co.

110 *o, p, q*. 3 animals, in spirits.
Australia. 59, 6, 30, 13. Presented by Dr. G. Bennett.

110 *r*. Animal, in spirit; female.
Australia. Presented by Prof. Owen.

110 *s*. Animal, in spirits; very young.
Ornithorhynchus paradoxus, *Owen, Phil. Trans.*
Australia. Zool. Soc. 55, 12, 24, 426.

735 *a*. Skeleton.
Australia. Dr. Mantell.

735 *d*. Skeleton (imperfect).
Australia. Zool. Soc.

735 *f*. Skeleton, young.
Australia. Presented by Professor Owen.

735 *g*. Skeleton.
Australia. Presented by Professor Owen.

735 *b*. Skull; male.
Australia. Sholaven. Presented by J. MacGillivray, Esq.

112 *a*. Animal stuffed. 13½ in ; male.
Van Diemen's Land. Presented by Gen. Hardwicke.

735 *c*. Skull.
Australia.

735 *e*. Skull.
Australia. Yarrell.

2. ECHIDNA, *Gray, Cat. Carniv. &c.* p. 394.

1. ECHIDNA ACULEATA, *Gray, Cat. Carniv. &c.* p. 394.
Echidnée, *Cuv. Oss. Foss.* v. t. xiii.

Head black. Spines much longer than the hair.

111 *g*. Animal, stuffed, not mounted, about 17 in.
Australia. Presented by the Haslar Hospital.

111 *c*. Animal, stuffed, about 16 in.
Australia. Presented by the Rev. F. W. Hope.

111 *h*. Animal, stuffed, not mounted, about 16 in.
Australia. Presented by the Haslar Hospital.

1017 *a*. Skeleton and skull, not mounted.
West Australia. Warwick.

1017 *b*. Skeleton. Skull wanting.
Australia. Zool. Soc.

1017 *c*. Skeleton.
Australia.

111 *e*. Animal, stuffed, about 21 in. Intermediate between *E. aculeata*
and *E. setosa*.
Australia. 58, 9, 2, 1. Presented by H. Farley, Esq.

111 *f*. Animal, stuffed, not mounted, about 16 in.
Australia. Presented by the Haslar Hospital.

111 *i*. Animal, stuffed, not mounted, about 15 in. Very hairy.
Australia. Presented by the Haslar Hospital.

111 *l*. Animal, not stuffed, about 12½ in.
Australia.

111 *k*. Animal, not stuffed, about 13 in. Very hairy.
Australia.

1006 *a*. Skeleton and skull, mounted.
Van Diemens' Land. Presented by Capt. Mangles.

1006 *b*. Skeleton, not mounted.
Van Diemen's Land. Zool. Soc.

2. ECHIDNA SETOSA.
Head pale. Abundance of hair between the spines.

112 *b*. Animal, stuffed, about 20 in.
Van Diemen's Land. 37, 4, 8, 102.

112 *d*. Animal, stuffed, about 17 in.
Van Diemen's Land. Gould. 41—1162.

112 c. Animal, stuffed, about 12½ in.
Van Diemen's Land. Gould. 41—1164.

1006 a. Skeleton and part of skin.
Van Diemen's Land. Captain Mangles.

111 b. Animal, in spirits, adult.
Queensland. 66, 7, 3, 1.   Presented by Sir D. Cooper, Bart.

111 d. Animal, in spirits, adult.
Australia, Port Stephens. 72, 11, 8, 1.   Pres. by Dr. G. Bennett.

111 j. Animal, in spirits.
Australia. 64, 12, 19, 3.

111 k. Very young fœtus, in spirits.
Australia. 65, 10, 1, 4.   Presented by Dr. Müller.

INDEX

TO THE GENERA AND SPECIES OF EDENTATA.

Order BELLUÆ.

Gray, Cat. Carniv. &c. p. 249.

Section 1. ORTHOGNATHA, *Gray, Cat. Carniv. &c.* p. 257.

The jaws of the usual shape; dental edge nearly straight, with the three kinds of teeth of the usual form and shape. Teats abdominal.

Sub-Section 1. GEOTHERIUM.

Skull and jaws tapering in front. Nostrils terminal. Eyes lateral. Terrestrial.

Sub-Order 1. NASUTA, *Gray, Cat. Carniv. &c.* p. 251.

Family 1. TAPIRIDÆ, *Gray, Cat. Carniv. &c.* p. 252.

Tribe 1. TAPIRINA, *Gray, Cat. Carniv. &c.* p. 253.

1. TAPIRUS, *Gray, Cat. Carniv. &c.* 254.

* *Tapirus.*

1. TAPIRUS TERRESTRIS, *Gray, Cat. Carniv. &c.* p. 254; *P. Z. S.* 1872, p. 484.

38 *a.* Animal, stuffed; half-grown.
S. America.   Presented by Mus. Roy. Coll. Surg.

31 *c.* Animal, stuffed; adult.
S. America. Zool. Soc.

38 *f.* Animal, stuffed; young.
Tapirus terrestris, *Gray, P. Z. S.* 1872, p. 492, t. xxii. f. 3.
S. America.

38 *b.* Animal, stuffed; young.
Cabai Elephantipede, *Geoff. Mus. Paris. Desm. Dict. Hist. Nat.* p. 503.
S. America.

38 *d.* Animal, stuffed; very young. 69, 3, 31, 9.
709 *l.* Skull of "*d.*" 72, 4, 11, 4.
Peru. Bartlett.

38 *e.* Animal, stuffed; very young.
S. America. 56, 1, 12, 1.

709 *d.* Skeleton.
Brazils. Becker.

709 *h.* Skeleton; young.
America.

709 *a.* Skull; adult.
America.

709 *c.* Skull; adult.
Brazils.   Presented by John Miers, Esq.

709 *b*. Skull; adult.
  Berbice.

709 *f*. Skull; half-grown.
  S. America.

709 *e*. Skull; young.
  Demerara.   Presented by Sir R. Schomburgk.

709 *g*. Skull; female.
  East Peru. Bartlett.

709 *i*. Skull; female.
  East Peru. Bartlett.

709 *k*. Skull.
  East Peru. Bartlett.

709 *m*. Skull.
  Surinam.

2. TAPIRUS LAURILLARDI, *Gray, Cat. Carniv. &c.* p. 256.
709 *g*. Skull.
  Tapirus Laurillardi, *Gray, Cat.* p. 256, f. 32, 33.
  Venezuela ? Brandt.

3. TAPIRUS ECUADORENSIS, *Gray, P. Z. S.* 1872, p. 492, t. xxii. f. 2.
a. Animal, stuffed ; young.
  Tapirus ecuadorensis, *Gray, P. Z. S.* 1872, p, 492, t. xxii. f. 2.
  Ecuador, Macas. 72, 1, 24, 13.

4. TAPIRUS ENIGMATICUS, *Gray, P. Z. S.* 1872, p. 490.
1577 *e*. Young, stuffed.
  Tapirus enigmaticus or T. leucogenys, jun., *Gray, P. Z. S.* 1872,
    p. 490, t. xxii. f. 1.
1577 *f*. Skull of "*a*." 72, 1, 24, 10.
  Ecuador, Sunia, upper part of Cordilleras.

** *Cinchaeus.*
5. TAPIRUS LEUCOGENYS, *Gray, P. Z. S.* 1872, p. 488, t. xxi.
1577 *a*. Adult male, stuffed.
  Gray, l. c. p. 491, f. (Skull).
  T. leucogenys, *Gray, P. Z. S.* 1872, p. 488, t. xxi. f. 1.
1577 *b*. Skull of " *a*."
  Ecuador, Cordilleras at Sunia and Asuay. 72, 1, 24, 3.

1577 *b*. Young female, stuffed.
  T. leucogenys, *Gray, P. Z. S.* 1872, p. 489, t. xxi. f. 2.
  Ecuador, Cordilleras at Sunia and Asuay.

1577 *a*. Animal, stuffed ; adult female.
1577 *a*. Skull of above.
  Ecuador, Asuay. 72, 1, 24, 8.

1577 *d.* Animal, stuffed; young.
1577 *d.* Skull of " *d.*"
  Ecuador, Asuay. 72, 1, 24, 7.

1577 *e.* Skin, unstuffed. Fur very long.
1577 *e.* Skeleton of " *e.*"
  Ecuador, Sunia. 72, 1, 24, 12.

1577 *c.* Skeleton.
  Ecuador, Sunia.

1577 *b.* Skull; adult male.
  Ecuador, Sunia.

2. RHINOCHŒRUS, *Gray, Cat. Carniv. &c.* p. 259.

  1. RHINOCHŒRUS SUMATRANUS, *Gray, Cat. Carniv. &c.* p. 259.

17 *a.* Animal, stuffed; adult.
17 *b.* Skeleton of " *a.*"
  Sumatra.

17 *c.* Animal, stuffed; young. With longitudinal white lines and
    spots.
  Sumatra. 58, 4, 28, 22.

17 *d.* Skeleton.
  R. malayanus, *Gerrard, List of Bones, B. M.*
  Sumatra. Zool. Soc.

710 *b.* Skull.
  Sumatra.

710 *c.* Skull.
  Sumatra. 51, 11, 10, 38.

  Tribe 2. ELASMOGNATHINA, *Gray, Cat. Carniv. &c.* p. 260.

3. ELASMOGNATHUS, *Gray, Cat. Carniv. &c.* p. 261.

  1. ELASMOGNATHUS BAIRDII, *Gray, Cat. Carniv. &c.* p. 261.

1500 *a.* Animal, stuffed; nearly adult.
  Panama.

1500 *e.* Animal, stuffed; nearly half-grown.
1500 *e.* Skull of " *b.*"
  Panama. 72, 7, 36, 1.

1500 *f.* Animal, stuffed; young; with longitudinal white lines and
    spots.
  T. Bairdii, *P. Z. S.* 1867.
  Panama. 71, 1, 10, 206.

1500 *c.* Skeleton.
  Panama. Salvin.

1500 *d.* Skeleton; young.
  Panama. Zool. Soc.

1500 a. Skull.
   Pauama.   Presented by Dr. Sclater.
1500 b. Skull.
   Pauama.

Sub-Order 2. SOLIDUNGULA, *Cat. Carniv. &c.* p. 262.

Family 2. EQUIDÆ, *Cat. Carniv. &c.* p. 262.

1. EQUUS, *Cat. Carniv. &c.* p. 263.

   1. Equus caballus, *Cat. Carniv. &c.* p. 263.

704 f. Skeleton.
   Domestic. Stevens' Sale-rooms.

704 g. Skeleton; bad.
   Domestic.

704 h. Skeleton. "Pony."
   Domestic. Stevens.

704 a. Skull.
   Domestic.

704 b. Skull.
   Domestic. 43, 3, 20, 22.

704 c. Skull. "English Horse."
   Domestic. India.

704 e. Skull and separated bones.
   Domestic.

704 i. Skull.
704 j. Skull.
704 k. Skull.
704 l. Skull.
704 m. Skull.
704 n. Skull.
   Domestic. Holland. Jeude.

2. ASINUS, *Gray, Cat. Carniv. &c.* p. 267.

   1. Asinus vulgaris, *Gray, Cat. Carniv. &c.* p. 268.

740 b. Skeleton; bad state.
   Domestic.

740 a. Skull. 46, 1, 1, 2.
740 d. Skull.
740 e. Skull. 58, 6, 9, 18.
   Domestic. England.

740 c. Skull. "Russian Ass."
   Domestic. Zool. Soc.

740 f. Skull.
   Domestic. Holland. Jeude.

2. ASINUS ONAGER, *Gray, Cat. Carniv. &c.* p. 269.

705 *b.* Skeleton; female.
India. Zool. Society.

705 *a.* Skull and bones of body.
India. Kutch.   Presented by the Earl of Derby.

3. ASINUS HEMIONUS, *Gray, Cat. Carniv. &c.* p. 271.

976 *a.* Animal, stuffed; male.
Thibet.   Presented by Lord Gifford.

976 *b.* Animal, stuffed; male.
Thibet.   Presented by Hon. E. India Co.

976 *c.* Animal, stuffed; 9 months old.
Asinus Kiang, *Layard.*
Mesopotamia.   Presented by Dr. Layard.

976 *e.* Skeleton.
Brandt.

976 *a.* Skull.
Thibet.   Presented by B. H. Hodgson, Esq.

976 *b.* Skull.
Gray, P. Z. S., 1839.
Thibet.   Presented by B. H. Hodgson, Esq.

976 *c.* Skull.
Gray, P. Z. S., 1839.
Thibet.   Presented by B. H. Hodgson, Esq.

976 *d.* Skull.
Thibet, N. of Ladakh.   Presented by Earl of Gifford.

976 *f.* Skull.
Thibet, Ladakh. Strachey.   Presented by Hon. E. India Co.

976 *g.* Skull.
Nepal.   Presented by B. H. Hodgson, Esq.

976 *h.* Skull.
Nepal.   Presented by B. H. Hodgson, Esq.

4. ASINUS QUAGGA, *Gray, Cat. Carniv. &c.* p. 274.

1449 *a.* Animal, stuffed; very bad state.
S. Africa. Zool. Soc.

1449 *a.* Skeleton; male.
S. Africa. Zool. Soc.

5. ASINUS BURCHELLII, *Gray, Cat. Carniv. &c.* p. 275.

854 *a.* Animal, stuffed; adult.
S. Africa. Zool. Soc.

854 *b.* Animal, stuffed; young.
S. Africa. Cape Museum.

854 *c*. Animal, stuffed.
 Born in the Zoological Gardens. 64, 6, 1, 2.

854 *b*. Skeleton.
 S. Africa.

854 *a*. Skull; female.
 S. Africa. Caffraria. Sundevall.

854 *c*. Skull; male.
 S. Africa.

 6. ASINUS ZEBRA, *Gray, Cat. Carniv. &c.* p. 276.

706 *e*. Animal, stuffed; adult.
 S. Africa. Zool. Soc.

706 *a*. Skeleton.
 S. Africa.

706 *d*. Skull.
 S. Africa.   Presented by the Earl of Derby.

706 *b*. Skull.
 S. Africa.

706 *g*. Animal, stuffed. Hybrid between Asinus Zebra and Asinus
 hemionus.
 Bred at Knowsley.

706 *h*. Animal, stuffed. Hybrid between Asinus Zebra and Asinus
 vulgaris.
 Bred at the Zoological Gardens.

706 *c*. Skull; male. Hybrid between Zebra and Ass.
 Domestic.

706 *e*. Skull. Hybrid between Zebra and Ass.
 Domestic. Zool. Soc.

 Sub-Order 3. LAMINUNGULA, *Gray, Cat. Carniv. &c.* p. 278.
 Family 1. HYRACIDÆ, *Gray, Cat. Carniv. &c.* p. 279.

Orbit with a small separate bone in front.

Tribe 1. HYRACINA.

Thick part of crown with a narrow 'sagittal crest, separating the
 temporal muscles. Orbits incomplete. The supra-occipital bone
 margining the occiput above. Blade-bones elongate, longer than
 broad. Coracoid process slightly developed.

 1. HYRAX, *Gray, Cat. Carniv. &c.* p. 283.
 ⁕ *Dorsal spot black. Diastema very narrow, shorter than the three
 first grinders.*

 1. HYRAX CAPENSIS, *Gray, Cat. Carniv. &c.* p. 285. *Cuv. Oss. Foss.*
 ii. p. 127, 14, f. i. ii. (Skeleton). *Blainv. Ost.* t. ii. Teeth and
 skull.

54 *a*. Animal, stuffed; adult.
S. Africa.   Presented by Gen. Hardwicke.

54 *b*. Animal, stuffed; adult.
S. Africa.

54 *c*. Animal, stuffed; adult.
S. Africa. 38, 3, 10.   Presented by G. Dodgson, Esq.

54 *d*. Animal, stuffed; adult.
S. Africa. Krauss.

54 *f*. Animal, stuffed; adult.
S. Africa. 45, 7, 3, 190. Shot at the top of a tree. Dr. Smith.

54 *e*. Animal, stuffed; young.
S. Africa. Verreaux. 38, 6, 9, 112.

54 *g*. Skin, not stuffed.
S. Africa. 58, 9, 12, 11. Mus. Zool. Soc.

54 *h*. Skin, not stuffed; with skull.
S. Africa. 47, 11, 2, 2.

54 *i*. Skin, not stuffed; young; with skull.
S. Africa. 43, 12, 7, 8. Verreaux.

724 *h*. Skeleton, bones separate. Interparietal bone quadrangular.
Cape of Good Hope. Zool. Society.

724 *g*. Skeleton; young; bones separate; no skull; blade-bone rather
    wide, ½ longer than broad.
S. Africa. Zool. Society.

724 *i*. Skeleton; young; bones separate. Interparietal bone quadran-
    gular.
S. Africa. Zool. Society. 64, 8, 17, 24.

724 *b*. Skeleton; young; bones separate; no skull.
S. Africa.

724 *j*. Skull; adult. Nose and back part injured.
S. Africa. Gerrard, jun.

724 *c*. Skull; half-grown; small. Interparietal square.
S. Africa.

724 *d*. Skull; small.
S. Africa.

※ *Dorsal streak yellow. Diastema narrow.*

2. HYRAX BURTONII.

120 *a*. Animal, stuffed; adult.
Hyrax syriacus, *Gray, List Mam.* 1843, p. 187.
N. Africa, Egypt.   Presented by J. Burton, Esq.

120 *b*. Animal; adult.
H. syriacus, *Gray, List Mam.* 1843, p. 187.
N. Africa, Egypt.   Presented by J. Burton, Esq.

120 *c*. Animal; young.
H. syriacus, *Gray, List. Mam.* 1843, p. 187.
N. Africa, Egypt.   Presented by J. Burton, Esq.

120 *f.* Animal, stuffed; young.
"Senegal." Parzudaki. 48, 1, 18, 28.

120 *d, e.* Animals, in spirits; very young.

725 *b.* Skull, imperfect behind.
Hyrax abyssinicus, "b," *Gerrard, Cat. Bones,* p. 284.
N. Africa, Egypt.   Presented by J. Burton, Esq.

725 *d.* Skull.
Abyssinia.

725 *c.* Skull; very old.
Africa.

3. HYRAX BRUCEI, *Gray, Cat. Carniv. &c.* p. 287.
Ashkok. *Bruce's Travels,* t.
Hyrax syriacus, *Blainv. Ostéog.* t. ii. (Skull and teeth).

57 *d.* Animal, stuffed; young.
Hyrax abyssinicus, *Rüppell, MS. in B. M.*
Abyssinia. Rüppell.

57 *b.* Animal, stuffed.
H. abyssinicus, *Rüppell, Gray, List Mam.* p. 187.
Abyssinia. Rüppell.

57 *c.* Skin, unstuffed; female.
Abyssinia, Mayen, Senafé Pass. 3500 ft. Blandford. 69, 10, 24, 86.

1535 *d.* Skin, unstuffed; female.
1535 *b.* Skull of "*d.*" Imperfect.
Abyssinia, Adgrat. 8000 ft. 69, 10, 24, 37.

1535 *e.* Skin, unstuffed; adult male.
Skull of "*e.*"
Abyssinia, Adgrat. 8000 ft. Blandford. 69, 12, 23, 2.

1535 *g.* Skin, unstuffed; adult male.
Skull of "*g.*" Abyssinia, Adgrat. 8000 ft. 69, 10, 24, 42.

1535 *h.* Skin, unstuffed; half-grown female. 69, 10, 24, 7, 28.
Skull of "*h.*" Crown broken. 69, 12, 23, 4.
Abyssinia, Adgrat. 8000 ft. 69, 10, 24, 28. (No. 886).

1535 *i.* Skin, unstuffed; half-grown, male.
Skull of "*i,*" complete.
Abyssinia, Agula, Tigré. Blandford. (No. 663). 69, 10, 24, 34.

1535 *j.* Skin, unstuffed. No skull.
Senafé, Tigré. 7500 ft. Blandford. (No. 289). 69, 10, 24, 39.

1535 *f.* Skin, unstuffed; young female.
Skull of "*f.*"
Abyssinia, Amerlee Bay. 69, 10, 29, 31.

1535 *k.* Animal, unstuffed; half-grown; with skull.
Abyssinia, Mayen, Senafé, Pass. 3500 ft. (No. 118). Blandford.
69, 10, 24, 45.

1535 *l.* Skin, unstuffed; adult female; with skull.
Abyssinia, Adgrat. 8600 ft. Blandford. (No. 786). 69, 10, 24, 40.

1535 *m.* Skin, unstuffed; adult male; with skull.
Abyssinia, Adigrat. 8000 ft. Blandford. (No. 784). 69, 10, 24, 32.

1535 *n.* Skin, unstuffed; young female; with skull.
Abyssinia, Antalo, Tigré. Blandford. (No. 643). 69, 10, 24, 35.

1535 *o.* Skin, unstuffed; half-grown male; with skull.
Abyssinia, Adigrat, Tigré. 8000 ft. Blandford. (No. 884).
69, 10, 24, 33.

1535 *p.* Skin, unstuffed; young male; with skull.
Abyssinia, Adigrat, 8000 ft. Blandford. (No. 777). 69, 10, 24, 4.

1535 *q.* Skin, unstuffed; young female.
Skull of " *q.*"
Abyssinia, Wadela plateau. 10,000 ft. Blandford. (No. 543).
69, 10, 29, 30.

1535 *r.* Skin, unstuffed; adult.
Skull of " *r.*"
Abyssinia, Samtora, Wadela plateau. Blandford. (No. 544).
69, 10, 24, 29.

1535 *a.* Skeleton; female. Interparietal bone half oblong.
Abyssinia, Adigrat. 8000 ft. 69, 10, 24, 44.
Skull; adult.
Abyssinia, ? Dongola. (Ehrenberg). Pres. by W. T. Blandford, Esq.

1535 *s.* Skull; female.
Abyssinia, Mai, Wahiz, Tigré. 8000 ft. Blandford. (No. 562).
69, 12, 23, 3.

1535 *t.* Skull; broken.
Abyssinia. Blandford.

1535 *u.* Flat skin, without feet.
1535 *u.* Skull and skeleton of "*u*"; nearly adult. Abnormal. Occipital
bone slightly produced above the margin. Upper jaw with a
conical canine-like tooth on each side behind the grinders.
Hyrax,'*Flower, P. Z. S.* 1869.
Abyssinia. 69, 12, 23, 1. Presented by W. T. Blandford, Esq.

4. HYRAX RUFESCENS.

1593 *a.* Animal, stuffed; adult male. Rufous, especially on rump.
Skull of " *a.*"
Abyssinia, Adigrat. 8000 ft. Blandford. (No. 776). 69, 10, 24, 41.

1593 *b.* Animal, stuffed; half-grown female.
Skull of " *b.*"
Abyssinia, Adigrat. 8000 ft. Blandford. (No. 780). 69, 10, 24, 38.

5. HYRAX ALPINI, *Gray, Cat. Carniv. &c.* p. 287.

1592 *a.* Animal, stuffed; adult.
Skull of " *a.*"
Hyrax Alpini, *Gray, Ann. and Mag. Nat. Hist.* ser. iii. i. p. 45.
N. Africa. " Abyssinia." Leadbeater. 43, 5, 15, 2.

6. HYRAX SINIATICUS, *Gray, Cat. Carniv. &c.* p. 288.

725 *a.* Animal, stuffed; adult; without skull.
Asia, Palestine. Tristram.

725 *b.* Animal, stuffed; young.
N. Africa? 55, 2, 19, 4.

725 *c.* Skull; adult (like *Blainv. Ost.* t. ii.)
N. Africa?

7. HYRAX FERRUGINEUS, *Gray, Cat. Carniv. &c.* p. 288. *Ann. and Mag. Nat. Hist.* 1869, iii. p. 242.

1591 *a.* Animal, stuffed.
1591 *a.* Skull, showing change of teeth.
Abyssinia. Jesse. 69, 2, 2, 1.

8. HYRAX IRRORATA, *Gray, Cat. Carniv. &c.* p. 280. *Ann. and Mag. Nat. Hist.* 1869, iii. p. 243.

1590 *c.* Animal, stuffed; adult.
Hyrax syriacus, *Brandt.*
Skull of "*c.*"
Abyssinia. Brandt. 45, 11, 1, 9.

1590 *b.* Animal, stuffed; adult. Dorsal stripe obscure.
Skull of "*b,*" complete.
Hyrax irrorata, *var.* leucogaster, *Gray, Cat.*
Abyssinia. Jesse. 69, 2, 2, 3.

1590 *a.* Animal, stuffed; adult.
a. Skull; adult. Imperfect behind.
Abyssinia. Jesse. 69, 2, 2, 2.

2. EUHYRAX, *Gray, Cat. Carniv. &c.* p. 289.

* *Dorsal spot large, black. Diastema wide, longer than the three first grinders.*

1. EUHYRAX ABYSSINICUS, *Gray, Cat. Carniv. &c.* p. 290.

705 *a.* Animal, stuffed; adult male.
Euhyrax abyssinicus, *Gray, Ann. & Mag. Nat. Hist.* 1868, p. 47.
Abyssinia, Ankober. Capt. Cornwallis Harris.

705 *b.* Animal, not stuffed; adult female.
Skull of "*b.*"
Euhyrax abyssinicus, *Gray, Ann. & Mag. Nat. Hist.* 1868, p. 47.
Abyssinia, Ankober. Capt. Cornwallis Harris.

724 *a.* Skeleton, adult; mounted.
Hyrax capensis, "*a,*" *Gerrard, Cat. Bones,* p. 283.
Cape of Good Hope. Zool. Soc.

** *Dorsal streak white, linear. Diastema very wide.*

2. EUHYRAX BOCAGEI.
Hyrax Bocagei, *Gray, Cat. Carniv. &c.* p. 289.

1515 a. Animal, stuffed.
1515 a. Skull of above. Interparietal bone triangular.
Hyrax Bocagei, *Gray, Cat.* p. 289.
Angola. 68, 12, 19, 3. Presented by Prof. T. V. B. du Bocage.

Tribe 2. DENDROHYRACINA.

The upper surface of the supra occipital bent backwards, forming part of the crown. Back of crown in adult animal flattened, separating the temporal muscles. Blade-bone short, broad, as broad as long. Orbits generally complete. Coracoid process of the blade-bone large, beak-like, and separated from the other bone by a cross suture.

3. DENDROHYRAX. *Gray, Cat. Carniv. &c.* p. 291.

* *Orbits entire, complete.*

1. DENDROHYRAX DORSALIS, *Gray, Cat. Carniv. &c.* p. 292.
Hyrax arboreus, *Blainv. Ostéog.* t. ii. (skull and teeth).
D. arboreus, *Blainv. Ostéog.* t. ii. (skull) ; not A. Smith.

1142 a. Animal, stuffed.
Hyrax dorsalis, *Fraser, P. Z. S.* 1852, p. 99.
Fernando Po. Fraser.

1142 d. Animal, stuffed; young. Dorsal spot large, white.
Ashantee. 72, 2, 22, 4.

1142 c. Animal, stuffed; young. Dorsal spot moderate.
W. Africa. 69, 5, 31, 2.

1142 f. Skin, unstuffed; young. Dorsal spot moderate.
Hyrax arboreus, *Verreaux.*
W. Africa. Verreaux. 52, 5, 1, 12.

1142 g. Animal, stuffed; young. Dorsal spot very small.
Ashantee. 72, 2, 22, 3.

1142 b. Skeleton, young.
W. Africa. 61, 10, 9, 3.

1142 h. Skeleton.
Fernando Po. 65, 5, 9, 9.

1142 a. Skull, adult.
" Fernando Po." Fraser.

725 a. Skull, wanting lower jaw; adult.
Hyrax abyssinicus, " c," *Gray, List Mam.* p. 187.
Ashantee. 42, 4, 26, 1. Presented by J. Reid, Esq.

724 f. Skull, very young; without lower jaw.
Hyrax capensis, young, *Gerrard, Cat. Bones,* p. 283.
Fernando Po.

2. DENDROHYRAX ARBOREUS, *Gray, Cat. Carniv. &c.* p. 292.

1586 *d.* Animal, unstuffed ; young.
S.E. Africa. 59, 5, 7, 20. Presented by Sir A. Smith.

1586 *b.* Animal, stuffed.
1586 *b.* Skull of above.
D. arboreus, *Gray, Ann. & Mag. Nat. Hist.* 1873, xi. p. 154.
S.E. Africa. 72, 10, 21, 2.

1586 *a.* Animal, stuffed.
1586 *a.* Skull of above.
D. arboreus, *Gray, Ann. & Mag. Nat. Hist.* 1873, xi. p. 154.
S.E. Africa. 72, 10, 21, 3.

1586 *c.* Skin, unstuffed.
1586 *c.* Skull of above.
S.E. Africa. 72, 10, 21, 1.

1586 *f.* Skin, not stuffed ; young. No white spot on back.
Hyrax capensis, *Verreaux.*
Cape of Good Hope. Verreaux. 43, 12, 7, 9.

*** *Orbits slightly incomplete behind.*

3. DENDROHYRAX BLAINVILLEI, *Gray, Cat. Carniv. &c.* p. 293.

724 *e.* Skull, adult.
H. capensis, "724 *e*," *Gerrard, Cat. Bones,* p. 283.
Heterohyrax Blainvillei, *Gray, l. c.* p. 293.
S. Africa. Zool. Soc. 58, 5.

4. DENDROHYRAX SEMICIRCULARIS, *Gray, Cat.* p. 284.

724 *h.* Skeleton, young.
Hyrax semicircularis, *Gray, l. c.*
S. Africa ? Zool. Gardens.

Blade-bone not quite so broad (3—4) as long, with a well-developed separate coracoid process. Orbit nearly united behind.

Sub-Order 4. NASICORNIA, *Gray, Cat. Carniv. &c.* p. 295.

Family 1. RHINOCEROTIDÆ, *Gray, Cat. Carniv. &c.* p. 300.

It is very difficult to determine to which species of *Rhinoceros* the horns in the Museum Collection belong, for two reasons. If purchased from dealers or from old collections no reliance can be placed upon the habitat assigned to them. Secondly, the horns of the same species seem to vary greatly in shape, and those from animals in confinement, and probably, therefore, also those that are wild, are considerably modified by external circumstances in shape and direction. For example, see account of animal with monstrous horn (*Sclater, P. Z. S.*) ; and therefore the separate horns in the collection are referred to the species with great doubt, unless they have been received from travellers who have obtained them themselves, and vouch for the accuracy of the habitat.

The Asiatic Rhinocerotes have the front of the nasal bone convex, produced, and more or less acute in front. The intermaxillaries in the skull of the very young animal are spongy and united together in front, with two rudimentary teeth on the hinder part of each side. In the older animals these teeth are more elongate, produced, and separate from each other in front, and supported by a more or less long process of the intermaxillary bone, which encases the upper and outer side of their hinder part. They have two teeth on each side, the hinder being the smallest; but in the older animals both these teeth drop out, and the front one is replaced by a large tooth, which eventually has a large flattened crown.

* One-horned Rhinocerotes.

Nose with one horn. Lower jaw with a pair of small cylindrical central cutting-teeth between the large lateral ones. The central cutting-teeth rarely absent in the adult skulls.

1. RHINOCEROS, *Gray, Cat. Carniv.* &c. p. 300.

In the Asiatic one-horned Rhinocerotes (*Rhinoceros*) there is a small cylindrical cutting-tooth on the inner side of the two lateral ones. These teeth are close to the inner side of the lateral ones in the skull of the fœtal animal; but they become separated from them as the front of the jaw dilates for the secretion of the permanent cutting-teeth, and when the larger lateral cutting-teeth are developed they are more compressed together. They are generally present; but there is a skull of *Rhinoceros javanicus* in the Museum (723 *a*) in which they are deficient, the inner sides of the large lateral cutting-teeth being very close together.

In the lower jaw of the skulls of very young animals there is a large conical cutting-tooth on each side in front. This tooth is very depressed, and has sharp edges on the side, and a half-ovate end. It becomes worn down and is replaced by a larger tooth, which becomes worn down on the upper surface so as to produce an elongated flat disk with an acute front. In the skulls of adult two-horned Asiatic Rhinocerotes (*Ceratorhinus*) these two middle cutting-teeth are wanting. I have never seen a very young skull of these animals.

1. RHINOCEROS JAVANICUS, *Gray, Cat. Carniv. &c.* p. 300.
R. sondaicus, *Horsfield*, Java, t. (animal).
S. *Müller, Verhandl.* t. xxxiii. (animal, male and female).
F. javanus, *Blainv. Ost.* t. i. ii. & vii.
R. sondaicus, *Cuv. Oss. Foss.* t. xvii. xviii (skeleton). *Blyth, Journ. Asiatic Soc. Bengal,* xxxi. t. 1 & 2 (skull).

723 *d.* Skeleton with skull; adult.
Java. Mus. Leyden.

723 *e.* Skeleton with skull; half-grown.
"R. sumatranus," Mus. Leyden.
Sumatra. Mus. Leyden. Franks.

723 *f.* Skeleton, imperfect.
R. javanicus, 3, *Gray, l. c.* p. 301.
Sumatra. Janson.

723 *a.* Skull, adult, without any lower central cutting-teeth.
Java?? Zool. Soc.

722 *h.* Skull, adult, with nasal bones cut off.
R. javanus, *Gray, Mus. Cat.* p. 301.
"R. unicornis," Zool. Soc.
Java? Zool. Soc.

2. RHINOCEROS UNICORNIS, *Gray, Cat. Carniv. &c.* p. 302; *Cuvier,
Oss. Foss.* ii. f. i.—iv.; *Blainv. Ost.* t. ii.
Indian rhinoceros, *Parsons, Phil. Trans.* 1742, t. 1 & 2.

88 *a.* Animal, stuffed.
India. Atkins' Menagerie.

722 *f.* Skeleton, adult.
India. Zool. Soc.

722 *g.* Skull, adult.
India. Zool. Soc.

722 *a.* Skull.
India.   Presented by C. Gascoin, Esq.

722 *b.* Skull, very young; perhaps just born.
R. unicornis, *Gray, Cat.* p. 303.
Nepal.   Presented by B. H. Hodgson, Esq.

722 *c.* Lower jaw, left side.
Nepal.   Presented by B. H. Hodgson, Esq.

722 *k.* Skull. Horn, 15 in.
Assam, Gompore.   Presented by Lieut-Col. Russell, Bart.

722 *o.* Skull, young. Front end of lower jaw broad, dilated. Front
cutting-teeth subcylindrical, far apart, showing permanent
cutting-teeth at base.
India.

722 *h.* Nose horn, 20 in.
India? (90 *b*).

722 *i.* Nose horn, 29 in.

722 *p.* Nose horn.
India? (125 *g*).

722 *j.* Nose horn, 10 in. From animal in confinement.
India? (89 *b*).

722 *b.* Nose horn, 8 in.
India. (87 *c*).

722 *m.* Nose horn, 6 in. Animal in confinement.
India. Cobbe. 37, 6, 10, 260.

722 *n.* Nose horn, 6 in. Animal in confinement.
India.

3. RHINOCEROS NASALIS, *Gray, Cat. Carniv. &c.* p. 305, f. 34, 35 (skull); *P. Z. S.* 1867, p. 1012, figs. 1 & 2. (skull).

723 *b.* Skull; nearly adult; wanting the intermaxillaries and the inner cutting-teeth, with very large lateral cutting-teeth.
"R. sondaicus, *Cuv.* Java," dealer.
Borneo?

723 *c.* Skull; nearly adult.
R. nasalis, *Gray, l. c.* p. 305, f. 34—35 (skull).
Borneo. Wright. 59, 8, 16, 1.

4. RHINOCEROS STENOCEPHALUS, *Gray, Cat. Carniv. &c.* p. 310, fig. 38—39 (skull); *P. Z. S.* 1867, p. 1018, f. 5, 6 (skull).

722 *e.* Skull; half-grown.
Rhinoceros stenocephalus, *Gray, Cat. Carniv. &c.* f. 38, 39.
Asia. Zool. Soc.

** *Two-horned Asiatic Rhinocerotes.*

Nose horns two, one behind the other. The lower cutting-teeth two, lateral. No intermediate ones in adult. Skin smooth.

2. CERATORHINUS, *Gray, Cat. Carniv. &c.* p. 313.

1. CERATORHINUS SUMATRANUS, *Gray, Cat. Carniv. &c.* p. 313.
R. lasiotes, *Sclater, P. Z. S.* 1872, p. 494, pl. xxiii.
Rhinoceros de Sumatra, *Cuv. Oss. Foss.* ii. 27, t. iv.
R. sumatrensis, *Blain. Ostéog.*
Sumatran rhinoceros, *Bell, Phil. Trans.* 1793, p. 3, t. ii., iii., iv. (copied *Cuv. Oss. Foss.* iii. p. 42, t. 78, f. 8).
R. sumatranus, *Blainv. Ostéog.* t. (skull). *Blyth, Journ. Asiatic Soc. of Bengal,* xxxi. 1862, p. 151, t. 3 (skull). *Müller, Verhand.* t. xxxv. (animal).

1461 *a.* Skull; adult.
Pegu. Theobald.

1461 *b.* Skull; nearly adult.
Sumatra. Raffles. Zool. Soc.

1461 *c.* "Hinder horn 13 in." Blyth.
Sumatra? (89 f.)

The figure of the skull, like the figure of the animal, attached to Mr. Bell's paper in the 'Philosophical Transactions' (vol. lxxviii. 1793, p. 3, t. ii.—iv.) well represents this species, and has well-developed cutting-teeth in the lower jaw, and the space between the condyles of the skull narrow, which is the character of this species.

Home's figure of the skeleton of the Sumatran rhinoceros (Phil. Trans. 1821, t. xxii.), from the skeleton now in the Royal College of Surgeons, better represents the height of the skull, but scarcely sufficiently shows the distinction between the two species.

The figure of *R. sumatrensis,* ♀, Blainv. Ostéog. t. ii., is not so high behind as the skulls of either of the species, and in other respects is not characteristic.

2. CERATORHINUS NIGER.

C. Sumatranus (part), *Gray, Cat. Carniv. &c.* p. 313.

R. Crossii, *Gray, Ann. and Mag. Nat. Hist.* 1872, x. p. 209 (not horn).

1588 *d*. Animal, stuffed.

C. niger, *Gray, Ann. and Mag. Nat. Hist.* 1873, xi. p. 357.
"Sumatra?" Franks.

1588 *c*. Animal, stuffed.

R. Crossii, *Gray, Ann. and Mag. Nat. Hist.* 1872, x. p. 209.

b. Skeleton of "*e*"; mounted.

C. niger, *Gray, Ann.* 1873.

R. sumatranus, *Sclater, Zool. Soc. Guide.*

Malacca. Zool. Soc. 72, 12, 31, 1.

1076 *a*. Front horn; adult. 32 in. Very slender.

R. Crossii, *Gray, P. Z. S.* 1854, p. 270, f. (horn).

Sumatra?

1076 *b* (9 *e*). Two horns on skin of head; young. Front horn 9 in.

Sumatra?

1076 *c*. Horn. Slender and much curved, but not so slender or curved as 1076 *a*.

Hab. unknown. 72, 6, 12, 1.

The British Museum purchased from the Zoological Society the body of "1588 *e*," which was obtained by Mr. William Jamrach at Singapore, and which was captured at Malacca in 1871. It is peculiar for having a very rough skin, the body being covered with thick black hair; the tail is comparatively long and thin.

Mr. Edward Gerrard, jun., has preserved and stuffed the skin, and prepared a very complete skeleton of the animal.

The skull is very different from those of the Sumatran rhinoceros (*R. sumatranus*, Raffles), collected by Sir Stamford Raffles and now in the British Museum and in that of the Royal College of Surgeons, and from the skull which was purchased of Mr. Theobald, and proves most distinctly that I was right in stating the animal, when alive, to be very distinct from the Sumatran rhinoceros described and figured by Bell in the 'Philosophical Transactions' for 1793, to which Sir Stamford Raffles gave the name of *R. sumatranus*, under which name the Malaccan rhinoceros was exhibited at the Zoological Gardens and mentioned in the list of accessions in the 'Proceedings of the Zoological Society;' and I see by the report that a paper on the details of its visceral anatomy has been read to the Society by Mr. Garrod.

There has for many years existed in the British Museum a stuffed skin of a young specimen, "1588 *d*," which was purchased of Mr. Franks, of Amsterdam, as the young Sumatran rhinoceros; but there is reason to believe that this specimen was from Singapore, the port of Malacca.

The skull of the Malaccan rhinoceros is very like that of the Sumatran one; but it is shorter and broader than that of *R. sumatranus*. The hole in the cheek for the passage of the large vessels is oblong, much larger, and nearer the margin of the nasal aperture;

while in the two skulls of *R. sumatranus* it is smaller, circular, and some distance from the margin of the aperture. The front edge of the intermaxillary bones is broader, rounded, and not compressed nor nearly so much produced as the front edge of the intermaxillary bone of the adult skull of *R. sumatranus*, nor so much as in the skull of the young animal of the same species, which is shorter and broader than in the adult. The grinders of the upper jaw are six in number, and appear broader than those of the adult *R. sumatranus*, but they occupy the same length.

The skull of the Malaccan rhinoceros is not so high behind as that of the adult Sumatran rhinoceros; and the space in the crown between the temporal muscles is flat, and much wider than that of the adult but not so aged Sumatran rhinoceros in the British Museum. The back end of the upper part of the occiput is not nearly so broad as that of the Sumatran rhinoceros.

The most striking difference is in the lower jaw. The condyles are further apart; indeed the whole jaw is wider; but the outer edge of the hinder angle is much more expanded. This latter peculiarity, as well as the form of the crown of the grinders in the upper jaw, may arise from the greater age of the specimen. The greatest peculiarity is that the front of the lower jaw is comparatively thin, expanded, and having neither teeth nor alveoli, nor, indeed, one may say, sufficient thickness to hold the large cutting-teeth usually found in the front of the lower jaw of this genus. The grinders are six on each side; that is to say, the front tooth on each side is retained, whereas it is shed from the skull of the adult but much less aged animal of *R. sumatranus* in the British Museum; and the grinders appear to differ in the form of their folds from those of the Sumatran species.

	niger. in.	sumatranus. in.
Length from tip of nose to occipital condyle of adult	21½	22
From front of intermaxillary to occipital condyle	20¼	21
From front edge to back edge of lower jaw	16½	17
Width at zygomatic arch	12	11
Width of hinder end of lower jaw	10⅞	9½
Width of upper part of lower jaw at end of tooth-line	7½	6¾
Height of back of skull	13	13½

It is very probable that the want of front teeth in the lower jaw may be an individual peculiarity produced by the age of the specimen; at least I do not think it safe to regard that peculiarity as specific without an examination of more specimens. The toothless front of the lower jaw is like that of the adult *Ceratotherium simus*, from S. Africa.

In the 'Annals and Magazine of Natural History,' 1872, x. p. 209, I referred to *R. Crossii* and thought it might be the same as *R. sumatranus* from Tavoy and Tenasserim, mentioned by Blyth, Journ. Asiat. Soc. Bengal, 1862, p. 156, who figures the skull and horns, and who identifies his animal with my *R. Crossii* (which was described from a pair of horns, P. Z. S. 1854), and has just informed me that it is the head of the small black rhinoceros with two horns.

Probably he is correct in thinking that the horn I figured as *R. Crossii* belongs to the same species as the skulls which he received from Tenasserim; but it is to be observed that I have never seen a skull of

H

the Tenasserim rhinoceros, and do not know whether it is the same as *C. sumatranus* from Sumatra or *C. niger* from Malacca, or whether it may be a distinct species. Therefore I think it best, until we receive skulls of the Tenasserim species, to give the Malaccan one a distinct name and call it *C. niger* (as the black colour at once distinguishes it from the greyish Sumatran species), more especially as some zoologists who admit the difference of the two species refer *R. Crossii*, of which we know nothing but the horn, to each of the species.

Mr. Blyth, in the 'Journal of the Asiatic Society of Bengal,' vol. xxxi. t. iii. f. 1, 2, 3, lithographs from photographs (which he has since given to me) three skulls of what he calls *R. sumatranus* from Tenasserim.

These skulls, according to the photographs, differ so much from each other that they do not afford materials for the determination of the question of the species to which the Tenasserim rhinoceros should be referred.

The photographs represent the skulls of animals of very different ages; but I cannot believe the difference between them depends solely on age, as the skull of the oldest (fig. 1) and of the youngest (fig. 3) agree in the shape of the occiput and in the upper surface not being produced behind, while the skull of the half-grown one (fig. 2) has the upper surface of the occiput very much produced backwards, and the occipital condyles not so prominent.

Mr. Blyth informs me that he believes the adult skull (t. iii. f. 1) is the skull of *R. Crossii*, which he thinks is *R. lasiotis*, and he believes that the two younger skulls (t. iii. f. 2 & 3) belong to the black rhinoceros. The youngest skull (t. iii. f. 3) has the skin of the head and horns attached to it in the Museum at Calcutta. But the form of the lower jaw in the two younger specimens do not agree with the lower jaw of *C. niger*, and therefore I should provisionally name them *C. Blythii*.

*** *African Rhinocerotes.*

The African Rhinocerotes have the intermaxillary bones small, laminar, situated on the front end of a bony plate, separated by a suture (which becomes obliterated in the older specimens), in the inner side of the front part of the maxillæ, and have a tooth on its edge, which generally falls out in the adult animal; hence usually described as having no intermaxillary cutting-teeth. The lower jaw of the young *R. bicornis* (1365 *b*) has a small cylindrical cutting-tooth on each side of the broad end of the jaw, which disappears in the older animals; and the breadth of the front of the jaw does not increase, and therefore becomes smaller compared with the size of the skull. In the skull of the fœtal specimen of *R. bicornis*, 8¼ in. long (1365 *h*), with the three grinders but partially developed, the intermaxillaries are cartilaginous, and show rudiments, or rather nuclei, of two teeth.

The lamina on the inside of the maxillæ of these African Rhinocerotes, bearing the intermaxillaries, is represented in the Asiatic Rhinocerotes by a broad portion of the inside of the maxillæ, which is marked by an external groove; but in these animals the broad intermaxilla is attached to the end of the maxillæ, as well as to the end of this defined part.

3. RHINASTER, *Gray, Cat. Carniv. &c.* p. 316.

* Rhinaster, *Gray, Cat. Carniv. &c.* p. 316.

1. RHINASTER BICORNIS, *Gray, Cat. Carniv. &c.* p. 316.
R. bicornis, *Smith, Ill. Zool. S. A.* t. ii. (animal).
R. bicorne, *Cuv. Oss. Foss.* ii. p. 29, t. iv. & xvi. *Blainv. Ost.* t.
iii. & iv.
R. africanus, *Harris, Wild Animals S. Africa,* t. xi.

1365 *c.* Animal, stuffed ; adult.
S. Africa. Presented by the Earl of Derby.

1365 *b.* Animal, stuffed ; half-grown.
S. Africa. Verreaux. Collected by Sir A. Smith.

1365 *a.* Animal, stuffed ; very young.
S. Africa. Collected by Sir A. Smith.

1365 *l.* Animal, stuffed.
1365 *l.* Skull.
Abyssinia, Bogos.

1365 *h.* Skin, unstuffed ; very young animal.
1365 *h.* Skull.
Abyssinia. 71, 11, 29, 5.

1365 *k.* Skin of head, stuffed, with rather slender conical elongate
horns ; female.
Skull of " *k.*"
Abyssinia, Anseba Valley. Pres. by the Bombay Government.

1365 *g.* Skeleton.
Abyssinia.

1365 *m.* Skull. Front horn thick, conical. Back horn thick, short.
Sennaar. Petherick.

1365 *e.* Horns, front 27 in. and back 8 in., conical, thick, with a
circular outline.
S. Africa. 60, 9, 29, 3.

1365 *i.* Front horn, 25 in. Very thick.
S. Africa ? (90 *a.*)

1365 *j.* Front horn. Very thick, cracked.
Abyssinia. 73, 2, 6, 1.

** Keitloa, *Gray, Cat. Carniv. &c.* p. 317.

2. RHINASTER KEITLOA, *Gray, Cat. Carniv. &c.* p. 317.
R. Keitloa, *Smith, Ill. Zool. S. Africa,* t. 1.
R. bicornis, *Camper, Act. Petrop.* 1777, ii. p. 193, t. iii., iv., v. (copied
Cuv. Oss. Foss. ii. t. iv. f. 5).

a. Animal, stuffed ; adult.
Rhinoceros Keitloa, *A. Smith, Ill. Zool. S. Africa,* t.
S. Africa. Verreaux.

1520 *a.* Skeleton ; female.
Abyssinia. Jesse.

1520 *d.* Front 17 in., and back horn 11 in.; half-grown.
S. Africa. 52, 12, 15, 12.

1520 *b.* Front horn, 23½ in.; adult.
S. Africa.

1520 *e.* Back horn, 23 in. Perhaps belonging to " *b.*"
S. Africa.

1520 *f.* Back horn, 12 in.
S. Africa.

1520 *h.* Front horn very slender, compressed at the tip, 42 in.
S. Africa.

1520 *i.* Front horn rather slender, very compressed at the end, 40 in.
S. Africa. (125 *a.*)

1365 *b.* Front horn, 17 in.; back, 5 in. Front horn conical, attenuated,
with a circular outline. Hinder horn short, compressed, flattened
on the sides.
R. bicornis, *b.* *Gerrard, Cat. of Bones,* p. 282.
S. Africa.

90 *b.* Hinder horn, 7¾ in.

4. CERATOTHERIUM, *Gray, Cat. Carniv. &c.* p. 319.

1. CERATOTHERIUM SIMUM, *Gray, Cat. Carniv. &c.* p. 319.

1003 *a.* Animal, stuffed; half-grown.
Rhinoceros simus, *A. Smith, Ill. Zool. S. Africa.*
S. Africa. Verreaux.

1003 *a.* Skull; adult.
1003 *b.* Skull; adult.
1003 *c.* Skull; young.
S. Africa.

1003 *d.* Skull; young.
S. Africa.

1167 *b.* Front horn, 57 in., very slender and nearly straight.
Rhinaster Oswellii, *b, Gerrard, Cat. of Bones,* p. 283.
S. Africa.

1003 *e.* Front horn elongate, 43 in.
S. Africa?

1003 *f.* Front horn, 36 in.; slender and circular at base.
S. Africa? (125 *b*).

1003 *g.* Front horn, 32 in.; rather compressed, thick.
S. Africa? (125 *f*).

1003 *h.* Front horn; thick.
S. Africa? (125 *d*).

2. CERATOTHERIUM OSWELLII, *Gray, Cat. Carniv. &c.* p. 322.

1167 *a.* Front and back horns. Front horn, 29 in., nearly straight;
back horn compressed.

Rhinoceros Oswellii, *Gray, P. Z. S.* 1854, p. 46, fig. (horns). *Andersson, Lake N'Gami,* p. 388, fig.
S. Africa, Lake N'Gami. Presented by Col. Thomas Steele.

1167 *c.* Front horn, 37 in.; nearly straight.
S. Africa.

1167 *d.* Front horn, 32 in.; curved back below and forwards at the tip.
Parsons, Phil. Trans. 1742—43, tab. 3, fig. 6 (horn).
S. Africa. Sir Hans Sloane's collection.

Sub-Order 5. SETIFERA, *Gray, Cat. Carniv. &c.* p. 325.

Section 1. HOMODONTINA.

Premolars permanent, forming with the molars a continuous series. Molars solid, tubercular.

Division 1. PSEUDOPERISSODACTYLA.

Hinder feet with three toes. The short external lateral toes of the hind feet wanting. America or Western Hemisphere. *Gray, Ann. & Mag. Nat. Hist.* 1873.

Family 1. DICOTYLIDÆ, *Gray, Cat. Carniv. &c.* p. 350. *Ann. & Mag. Nat. Hist.* 1873.

Cutting-teeth $\frac{2-2}{3-3}$ Premolars $\frac{3-3}{3-3}$ Upper canines bent down; front upper margin of the sheath of the upper teeth more or less thickened in the margin.

1. NOTOPHORUS, *Gray, Cat. Carniv. &c.* p. 350.

Groove of vessel over eye curved to the margin, and then bent back over the canines and continued to the end of the nose. The very young animal pale brown, with a paler collar.

1. NOTOPHORUS TORQUATUS, *Gray, Cat. Carniv. &c.* p. 351.
Sus torquatus, *Blainv. Ost.* t. iii. (skeleton).

55 *a.* Animal, stuffed; adult; bad condition.
Brazils.

55 *b.* Animal, stuffed.
Brazils. 41, 12, 20, 1. Presented by A. Cross, Esq.

55 *e.* Animal, stuffed; half-grown; blackish.
Brazils. 53, 8, 29, 23.

55 *f.* Animal, stuffed; half-grown; blackish.
Brazils. 44, 7, 4, 6.

55 *d.* Animal, stuffed; very young; pale brown, with interspersed black hairs, more abundant on the middle of the back. Collar and under side of neck and body pale.
"Sus de Mozambique." Verreaux.
Brazils. 60, 2, 11, 15.

55 *c*. Animal, stuffed.
Skull of " *e*."
 Brazils. 42, 5, 16, 7.

55 *l*. Animal, stuffed ; young.
 Brazils. 43, 9.

55 *g*. Animal, unstuffed ; male ; large size, with skull.
 Brazils. 46, 11, 29, 9.

55 *h*. Animal, unstuffed ; brownish, with skull.
 Brazils. 41, 6, 1, 29.

55 *i*. Animal, unstuffed ; with skull.
 Costa Rica. Salvin. 65, 5, 18, 34.

55 *j*. Animal, unstuffed ; without skull.
 Costa Rica. Salvin. 65, 5, 18, 35.

720 *h*. Skull ; adult.
 S. America. 67, 4, 12, 207.

720 *i*. Skull ; adult male. Lower canines longitudinally grooved on
   the outer side.
 S. America.

55 *c*. Skull ; skin destroyed.
 Brazils. Mus. Leyden.

55 *k*. Skull, from skin in bad state.
 Brazils. Haslar Hospital.   Presented by Dr. Armstrong, R.N.

720 *e*. Skeleton ; adult male.
 Brazils. Zool. Society.

720 *g*. Skeleton, mounted.
 Brazils. Zool. Society.

720 *a*. Skull ; adult male.
 Brazils. 43, 5, 16, 8.

720 *b*. Skull ; adult male.
 Brazils, Organ Mountains. 38, 4, 16, 86.

720 *c*. Skull ; adult, with caries from confinement.
 S. America. Zool. Society. 50, 11, 22, 49.

720 *f*. Skull ; young, wanting three canines.
 Brazils. Zool. Society.

720 *d*. Skull ; adult male.
 Brazils. Zool. Society.

2. DICOTYLES, *Gray, Cat. Carniv. &c.* p. 351.

 Groove of vessel over orbits simple, only continued to the keel of the
   zygomatic arch. Young animal pale brown, pale below, without
   white tips or chin.

 1. DICOTYLES LABIATUS, *Gray, Cat. Carniv. &c.* p. 352.
 Dicotyles torquatus, *Blainv. Ostéog.* t. v. (skull).

152 *a*. Animal, stuffed ; adult.
 Brazils. 42, 8, 17, 1.

152 c. Animal, unstuffed; adult male. Skull enclosed.
Brazils, Bahia. Brandt. 43, 9, 27, 4.

153 d. Animal, stuffed; young.
721 a. Skull of " d," without lower jaw. Canines of both sets present.
Brazils. 45, 2, 11, 65.

152 b. Animal, stuffed; very young, pale brown. Lips and under jaw
whitish. Dorsal streak indistinct.
Prinz. Max., Abbildungen, t.
Brazils.

152 e. Animal, unstuffed; adult male, with skull.
Brazils. 46, 6, 1, 27.

152 f. Animal, unstuffed; very young, brown. Dorsal streak broad,
well marked.
Brazils. 43, 9, 14, 17.

721 c. Skeleton; male.
Brazils. Becker. 47, 4, 6, 9.

721 d. Skeleton; female.
Brazils. Becker. 47, 4, 6, 8.

721 f. Skeleton; adult male.
Brazils. Zool. Society. 61, 4, 8, 2.

721 e. Skull; male, very large. Wanting upper canines.
Brazils. 51, 8, 27, 76.

721 b. Skull; young.
Brazils. 43, 2, 20, 6.

Division 2. ARTIODACTYLA.

*Fore and hind feet with four toes, the lateral toes of each side
much shorter.* Eastern Hemisphere or Europe, Asia and Africa.
Gray, Ann. & Mag. Nat. Hist. 1873.

Family 1. SUIDÆ, *Gray, Cat. Carniv. &c.* p. 327. *Ann. & Mag.
Nat. Hist.* 1873, p.

Head conical. The upper canines of the males elongate and more
or less recurved, and enclosed in a bony sheath at the base.
Teeth 40—44. Cutting-teeth $\frac{3-3}{3-3}$ Premolars $\frac{3-3}{3-3}$ or $\frac{4-4}{4-4}$
Tail elongate. Teats ten, rarely eight. Skull, with the sides
of the nose in front of the orbit more or less deeply concave.
The males have thick large canines, and a longitudinal ridge over
the sheath over its base, and especially if they have large thick
canine teeth in the lower jaw, the sides of the jaw are dilated to
a little behind the canines; and in the species or individuals
that have the largest canines the front of the jaw is dilated, and
in this respect somewhat approaches the form of the lower jaw
of *Phacochœrus*.

Tribe 1. POTAMOCHŒRINA, *Gray, Ann. & Mag. Nat. Hist.* 1878.

Ears elongate, attenuated and pencilled at the end. The concavity in the skull in front of the orbits without any ridge on the lower part from the front of the zygomatic arch. The sheath of the upper canines expanded; of the males largest, and with a ridge across the upper surface; of the females, often slightly bent up at the outer margin.

1. POTAMOCHŒRUS, *Gray, Cat. Carniv. &c.* p. 340.

Zygomatic arch of the skull thick, broad, prominent and convex externally. The skull of the female is much like the skull of a young male. The zygomatic arch is slighter and less convex than in the male.

1. POTAMOCHŒRUS AFRICANUS, *Gray, Cat. Carniv. &c.* p. 341.
Sus Africanus, *Blainv. Ost.* t. viii. f. 1.
Sus larvatus, *Blainv. Ost.* t. v.

87 *a.* Animal, stuffed; adult.
S. Africa.

88 *c.* Animal, stuffed; adult.
S. Africa.

87 *d.* Animal; half-grown.
S. Africa.

1364 *b.* Animal, unstuffed; in very bad state.
1364 *b.* Skeleton.
S. Africa. 62, 3, 30, 3.

87 *f.* Animal, stuffed; very young, with skull. Brown; back blackish, with narrow interrupted pale stripes.
S. Africa, Cape of Good Hope. Verreaux. 43, 12, 7, 20.

1364 *b.* Skeleton.
S. Africa. Zool. Society.

1364 *a.* Skull; male.
Sus larvatus, *Blainv. Ostéog.* t. v.
S. Africa. 51, 5, 5, 3.

1364 *c.* Skull; male.
Cape of Good Hope. Jeude. 67, 4, 12, 220.

714 *b.* Skull; female, without lower jaw. The sunken plane before the orbits indistinctly defined and shelving off behind.
W. Africa. Baikie. 64, 7, 16, 8.

2. POTAMOCHŒRUS PORCUS, *Gray, Cat. Carniv. &c.* p. 342.

1363 *a.* Animal, stuffed; adult male.
1363 *a.* Skeleton and skull of " *a.*"
P. penicillatus, *Sclater, P. Z. S.* 1860, p. 361, fig. (skull).
Cameroons River. Zool. Society.

1363 *f*. Animal, unstuffed; very young, with skull. Pale brown, back
and sides of head with a few indistinct black streaks.
S. penicillatus, *P. Z. S.* 1861, p. 62, pl. 11.
West Africa. Born at the Zool. Gardens. 57, 8, 3, 1.

1363 *g*. Animal, stuffed; very young, rather pale. 56, 12, 30, 7.
1363 *h*. Animal, stuffed; very young, rather pale. 57, 3, 5, 1.
1363 *i*. Animal, stuffed; very young, blackish. 67, 10, 5, 21.
W. Africa. Born at the Zool. Gardens.

1363 *e*. Skin, unstuffed; wanting much hair.
Potamochœrus penicillatus, *Du Chaillu, Travels,* fig.
1363 *e*. Skull of "*e*."
Gaboon. 71, 5, 27, 6.   Presented by M. du Chaillu.

1363 *b*. Skeleton; female. Skull with only a very slight marginal
keel on the sheath of the upper canines.
Cameroons River. Zool. Soc.

1363 *c*. Skull; male. Process of sheath of canine cut off. Wanting
canines and cutting-teeth.
Interior Africa. Baikie. 65, 5, 3, 4.

1363 *d*. Skull. Wanting cutting-teeth and upper canines.
Interior Africa. Baikie. 65, 5, 3, 4.

Tribe 2. SUINA, *Gray, Cat. Carniv. &c.* p. 328; *Ann. & Mag. Nat.
Hist.* 1873, June.

Ears rounded in the domestic, elongate, drooping, not pencilled at
the end. The concavity in the skull in front of the orbits with
a ridge on the lower part, from the front of the zygomatic arch.

Cutting-teeth $\frac{6-6}{6-6}$ The front grinders close on the back of

the upper canines, which in males are free and bent outwards
and upwards. The sheath of the upper canines of the males
spread out, with a ridge or crest across their upper surface, and
of the females often slightly bent up at the end.

1. Wild Swine. Ears moderate, hairy.

1. EUHYS, *Gray, Cat. Carniv. &c.* p. 339; *Ann. & Mag. Nat. Hist.*
1873.

Head elongate, twice as long as high at the occiput. Cheeks and
throat covered with long projecting hairs. Lower canines of the males
elongate, slender, convex on the sides and rounded in front. The front
false grinders near the base of the canines separated from the other
grinders by a rather broad diastema. Sheath of the upper canines in
the males with an elongated ridge, which has a straight top.

1. EUHYS BARBATUS.
Sus barbatus, *Gray, Cat. Carniv. &c.* p. 339. *Müller, Verhand.* i.
p. 42, t. 30; female animal, t. 31, and skull of female.

I

712 *a.* Skull; adult **male.** Wanting cutting-teeth and some canines.
S. barbatus, *Müller, Verh.* t. xxxi., t. 4 & 5.
Borneo. Presented by J. Brooke, Esq.

712 *b.* Skull; young, female. Lower jaw broken.
Sus timorensis, *Müller, Verhandl.*
Borneo. 47, 5, 11, 1. Presented by Sir E. Belcher.

2. AULACOCHŒRUS, *Gray, Ann. & Mag. Nat. Hist.* 1873.

Head conical, about one and a half times as long as high at the occiput. Male with a very high keel across the base of the sheath of the upper canines, which is thicker behind. The lower canines triangular, flat on the sides and keeled in front.

This genus has some alliance, in the large size of the sheath of the upper canines of the males, with *Potamochœrus.*

1. AULACOCHŒRUS VITTATUS.

Sus vittatus, *Gray, Cat. Carniv. &c.* p. 331. *Müller, Verhand.* i. t. 29 (animal), t. 32, f. 5 (skull, male). *Blainv. Ostéog.* t. v. (skull).

Head, body and legs covered with black bristles. Bristles of forehead and neck white-tipped. Streak round angle of mouth and along the lower part of the cheek white.

1362 *a.* Animal, stuffed. Skull enclosed.
" Sus verrucosus," *Franks.*
"New Guinea." Franks. 43, 12, 27, 1.

1362 *c.* Skull; adult male.
Sus vittatus, *Wallace.*
Amboyna. Wallace. 59, 4, 6, 6.

1362 *d.* Skull; adult male.
Sus vittatus, *Wallace.*
Amboyna. Wallace. 59, 4, 6, 5.

1362 *h.* Skull; old male.
Batchian. Wallace. 61, 12, 11, 26.

1362 *g.* Skull; male. Wanting one upper grinder.
Java? Mus. Utrecht. 67, 4, 12, 213.

1362 *f.* Skull; female. Wanting a few cutting-teeth.
Java? Mus. Utrecht. 67, 4, 12, 213.

3. DASYCHŒRUS, *Gray, Ann. & Mag. Nat. Hist.* 1873, p.

Head elongate, conical, more than one and a half times the length of its height at the occiput. Nose with a large flat-topped wart on each side, over the angle of the mouth, with a tuft of elongate pale bristles on the lower part of each cheek.

Males with a short rounded compressed ridge near the sheath of the upper canines. Lower canines triangular, flat on the outer side and keeled in front.

1. DASYCHŒRUS VERRUCOSUS.

Sus verrucosus, *Gray, Cat. Carniv. &c.* p. 330. *Müller, Verhand.*
t. p. 107—t. xxviii. (animal), t. xxxii. (skull, male).

Head nearly twice as long as high. Black, under side and front of
thighs pale.

712 *c.* Skull; adult male. Length 16 in. No upper canines. Im-
pression on right side of orbits narrowed behind. Upper
canines lost.

Müller, Verhand. t. iv. f. 3 & 4.
Java. 55, 4, 2, 3.

712 *d.* Skull; adult male. Lower jaw broken behind. Wanting some
of the cutting-teeth.

S. verrucosus, *var.* ceramicus, *Gray, Cat. Carniv. &c.* p. 330.
Ceram. Wallace. 55, 4, 14, 2.

712 *f.* Skull; adult male, perfect.
Sus larvatus, *Jeude.*
Java. 67, 4, 12, 211.

712 *i.* Skull; adult male, perfect.
Java. Mus. Utrecht. 67, 4, 12, 212.

136 *a.* Skull; adult male, rather broken in the nose, wanting most of
the teeth.

S. vittatus, *Wallace.*
Borneo.

712 *g.* Skull; adult male. Wanting the upper and some of the lower
cutting-teeth.
Java. 67, 4, 12, 216.

712 *h.* Skull; adult male. Wanting upper canines and some grinders.
Java. 67, 4, 12, 214.

2. DASYCHŒRUS CELEBENSIS.

Sus celebensis, *Gray, Cat. Carniv. &c.* p. 331. *Müller, Verhand.* i.
t. 28 (animal and skull).

The head about one and a half times as long as high at the occiput.
Animal black above and below.

1596 *a.* Animal, stuffed, not in good state, black.

Skull of "*a*," wanting upper canines and some other teeth.
Celebes. Mus. Leyden. 47, 5, 10, 2.

1596 *b.* Animal, stuffed; young, pale brown.
Celebes. Mus. Leyden. 48, 12, 27, 2.

The adult specimen has very long black bristles, especially on the
front of the back. The head has a round distinct wart on the upper
part of the cheeks, in a line with the angle of the mouth. The ears
are small and ovate. There is a tuft of long yellow bristles on the
cheeks under each eye.

The young is covered with reddish brown very long bristles, which
are long on the crown of the head and at the back of the lower part of
the cheeks. Hoofs brownish. Ears ovate, short. The warts on the
side of the nose are not at all apparent in the young stuffed specimens.

4. SUS, *Gray, Cat. Carniv. &c.* p. 329 (part); *Ann. & Mag. Nat. Hist.* 1873.

Head conical, about one and a half times as long as high at the condyles, without any, or only a very small wart on the side of the head. Ears ovate. The upper canines of the males recurved, with a more or less large keeled ridge across the sheath at their base. Lower canines of the males triangular, flat on the outer side and keeled in front.

1. SUS PAPUENSIS, *Lesson, Voy. Coquille,* t. viii. (animal and skull, pretty, but very badly represents the species).
Porcula papuensis, *Gray, Cat. Carniv. &c.* p. 339.

949 *a.* Animal, stuffed; adult female?
Sus papuensis, *Gray, List Mam.* 1843, p. 185.
New Guinea, 1843.   Presented by the **Earl of Derby.**

949 *b.* Animal, stuffed; young male.
949 *b.* Skull of above. 50, 7, 20, 134.
South Coast of New Guinea, Touton Island. 50, 9, 6, 15.
                                          Presented by Capt. Stanley.

The adult specimen received from the Earl of Derby lived long in confinement, and was sent as a male, but it has no external appearance of so being, and the canines are small and scarcely exposed. The head is without any warts on the cheeks, the ears are ovate and moderate. The tail is elongate, slender, naked except for a few bristly hairs at the end. The skin is nakedish, with scattered very slender bristles, which are longer on the back and sides, but longest and most abundant at the front of the shoulders and at the outside of the upper arms, where they are blacker. There are numerous hairs on the inner side of the front of the fore legs; a number of very short erect hairs over the upper part of the nose and front of the chin, where they are longer, making in fact a blackish muzzle, which is also to be seen in the young specimen. Length to base of tail, 59 in. Tail, 12 in.

The young specimen has much more abundant and longer hair, pale yellowish brown, the longer black hair forming a broad central vertebral line and three indistinct lines on the sides, and on the sides of the head. The skull of this animal, which is also said to be a male, but exhibits no external mark of being so, is evidently that of a young animal, with the hinder grinders in each jaw not developed, and indeed only slightly marked. The central cutting-teeth are small, lobed at the end. The hinder lateral cutting-teeth are rather elongate and slightly bent outwards.

It appears to me rather to belong to the genus *Sus* than *Porcula.*

2. SUS TIMORENSIS, *Gray, Cat. Carniv. &c.* p. 335. *Müller, Verhand.* i. t. 31, f. 1—3 (animal and skull, female.)
Head and body covered with very long black bristles, which are white at the end. Cheeks not warty.

1501 *c.* Skull; young male?
Wild pig, *Wallace.*
Macassar. Wallace. 59, 6, 4, 4.

1501 *b*. Skull ; male, rather larger than "*c*."
Wild pig, *Wallace.*
Ternate. Wallace. 61, 12, 11, 25.

1501 *a*. Skull; young female ? Changing teeth.
Sus timorensis, *Müller, Verhand.* t. 31, f. 2, 3 ?
" Young female *Babirussa*," *Wallace.*
Wallace.

3. SUS LEUCOMYSTAX, *Gray, Cat. Carniv. &c.* p. 333. *Temminck,*
Faun. Java, t. 20 (animal and skull, female). *Swinhoe, P. Z. S.*
1870, p. 640, figs. 1, 2 (head).
Animal black. Streak from angle of mouth, throat, under side of
neck, belly and inside of legs white ; female. Hoofs black.

1595 *a*. Skin, unstuffed ; male.
Sus leucomystax, *Swinhoe, P. Z. S.* 1870, p. 640, p. 1, 2 (head).
1595 *b*. Skull of "*a*." Canines elongate.
Sus taivanus, *Swinhoe, P. Z. S.* 1864, p. 382, 1870, p. 419.
Shanghai. Swinhoe. 70, 2, 10, 38.
May not this be the old boar of *Sus taivanus ?* It is doubtful
whether it is the same as the one described by Temminck.

4. SUS TAIVANUS, *Swinhoe, P. Z. S.* 1864, p. 383, 1870, p. 641,
f. 3, 4 (head).
Porcula taivana, *Swinhoe, P. Z. S.* 1862, p. 360.
Ears small, rounded. Hoofs black.

1594 *a*. Animal, stuffed ; young female ?
Sus taivanus, *Swinhoe, P. Z. S.* 1870, p. 642, fig. 2, 3 (head and
teeth).
1594 *a*. Skull of "*a* "; female ?
Formosa. Swinhoe. 69, 2, 10, 88.

1594 *b*. Skin, unstuffed; young female. Reddish ; sides and neck and
under part of body whitish.
Formosa. Swinhoe. 72, 2, 10, 41.

1594 *h*. Skull ; male.
Formosa. Swinhoe. 69, 2, 10, 85.

1594 *i*. Skull ; female.
Formosa. Swinhoe. 69, 2, 10, 87.

1594 *j*. Skull ; female.
Formosa. Swinhoe. 69, 2, 10, 84.

1594 *k*. Skull ; young female.
Formosa. Swinhoe. 69, 2, 10, 86.

1594 *l*. Skull, without lower jaw ; young. Back molars not developed.
Wild pig, *Collingwood.*
Formosa, S.W. part. 68, 10, 9, 1. Pres. by R. T. Collingwood, Esq.

1595 *m*. Lower jaw of a very old animal ; female. Grinders much
worn. Sent as belonging to "*l*."
Wild pig, *Collingwood.*
Formosa, S.W. part. 68, 10, 9, 1. Pres. by R. T. Collingwood, Esq.

A. Domestic varieties, with moderate bristles.

1594 c. Skin, unstuffed; young female, red.
"Red pig of Formosan Savages," *Swinhoe*.
Formosa. Swinhoe. **72, 2, 10, 42.**

1594 d. Skin, unstuffed; young, red.
"Red pig of Savages," *Swinhoe*.
1594 d. Skull of " *d.*"
Formosa. Swinhoe.

1594 e. Skin, unstuffed; young, without skull. Blackish brown, with
longer black hairs. Ears moderate, rather acute.
Formosa. Swinhoe. **70, 2, 10, 45.**

1594 h. Animal, stuffed. Black, with white feet and white on the sides,
the two sides being rather different. Ears small.
1594 h. Skull of " 1594 *h*," old male, with recurved canines; two lateral.
Hinder **upper** cutting-teeth absent, and the alveola on one side
filled up.
Formosa. Zool. Gardens.

B. Domestic variety, with abundant long hair. Bred by the Formosan
savages of the Eastern coast.

1594 f. Skin, unstuffed; young. No skull. White, varied with
reddish brown hairs on the upper part of the head.
East coast of Formosa. Swinhoe. **70, 2, 10, 43.**

1594 g. Skin, unstuffed; young, without skull. Black, with a white
streak up the forehead, the four feet and the belly white.
"Chefoo Pig," *Swinhoe*.
Amoy, June, 1867. Swinhoe. **70, 2, 10, 44.**

These pigs, bred by the savages of Formosa, differ from European
domestic pigs in having moderate-sized rounded ears with a small
acute point at the tip.

5. SUS MYSTACEUS, *Gray, Ann. & Mag. Nat. Hist.* 1873.
Brown, with scattered black bristles, which are more abundant on
the muzzle, forehead, sides of cheeks and sides of the front part of the
body. Crest and hinder part of body browner. Streak on each side of
nose over angle of mouth, elongate. Whiskers on black cheeks, gullet,
throat, chest, front of shoulders, the throat, thighs, and under side of
body whitish. Skull, concavity on side broad and deep, only separated
from the orbits by a very narrow ridge. The sheath of the upper
canines with a keeled ridge and convex on the outside of it.

712 *e.* Animal, stuffed; male.
712 *e.* Skeleton of above.
Sus verrucosus. "Javan Pig." Zool. Gardens. Not Müller.
Java? **62, 1, 22, 2.** Zool. Soc.

6. SUS CRISTATUS, *Gray, Cat. Carniv. &c.* p. 338.
Sus indicus, *Gray, List Mam.* p. 185. *Blainv. Ostéog.* t. v. (skull).
Crest over base of upper canines in males, moderate, rounded above.
Hoofs white, rarely black varied.

716 *a*. Animal, unstuffed; very young.
Nepal. Presented by B. H. Hodgson, Esq.

715 *p*. Animal, stuffed ; very young.
Nepal. 43, 1, 12, 90. Presented by B. H. Hodgson, Esq.

716 *t*. Skull ; adult male.
"Borner."
India, Nepal. 58, 6, 24, 123. Presented by B. H. Hodgson, Esq.

716 *b*. Skull ; adult male.
India. Presented by Gen. Hardwicke.

716 *y*. Skull ; adult male. Wants upper canines.
India. 57, 1, 4, 22. Presented by G. H. Money, Esq.

716 *x*. Skull ; adult male. Wants upper canines.
Crests of canines rather large.
India. Zool. Society. 58, 5, 4, 37.

716 *l*. Skull ; adult male.
Sus indicus, *Oldham*.
Nepal, Tarai. 56, 5, 6, 57. Presented by Prof. Oldham.

716 *e*. Skull ; adult male.
"Wild boar of the plains," *Hodgson*.
Nepal. 45, 1, 8, 87. Presented by B. H. Hodgson, Esq.

716 *d*. Skull ; adult male.
Wild boar of the plains.
Nepal. 45, 1, 8, 88. Presented by B. H. Hodgson, Esq.

716 *a*. Skull ; adult male. Very large tusks.
India. 41, 1, 12, 4.

716 *u*. Skull ; adult male.
"Marquis," *Hodgson*. Sus bengalensis, *Blyth*.
Nepal. 58, 6, 24, 123. Presented by B. H. Hodgson, Esq.

716 *p*. Skull ; adult male.
"Bilmarecah, *S. indicus*."
India. 52, 11, 12, 1.

716 *k*. Skull ; adult male.
"*Sus babirusa*."
Malabar. 38, 3, 13, 49.

716 *g*. Skull ; adult male.
"Marquis," *Hodgson*.
Nepal. Presented by B. H. Hodgson, Esq.

716 *o*. Skull ; adult male. Nose elongate. Wants canines.
India. Presented by Sir John Boileau.

716 *f*. Skull ; adult male.
Nepal. 45, 1, 8, 86. Presented by B. H. Hodgson, Esq.

716 *c*. Skull ; young male ?
"Wild boar of the plains," *Hodgson*.
Nepal, Tarai. 45, 1, 8, 89. Presented by B. H. Hodgson, Esq.

716 *w*. Skull ; young male, broken.
Sus bengalensis, *Blyth*.
Nepal, Tarai. 58, 6, 24, 132. Presented by B. H. Hodgson, Esq.

716 *m*. Skull, young female.
 Nepal, Tarai. 56, 5, 6, 58.   Presented by Prof. Oldham.

716 *q*. Skull; young female, imperfect.
 Nepal. 58, 6, 24, 125.   Presented by B. H. Hodgson, Esq.

716 *v*. Skull; nearly adult female.
 Sus affinis, *Gray, Cat. Osteol.* p. 71, 1847.
 Neilgherries. 38, 3, 13, 68.

716 *n*. Skull; adult female.
 India. 56, 5, 6, 59.   Presented by Prof. Oldham.

716 *j*. Skull; very young. With milk-teeth.
 Nepal. 45, 1, 8, 92.   Presented by B. H. Hodgson, Esq.

716 *i*. Skull; very young. With milk-teeth.
 Nepal. 45, 1, 8, 91.   Presented by B. H. Hodgson, Esq.

7. SUS ANDAMANENSIS, *Gray, Cat. Carniv. &c.* p. 336.

1497 *c*. Skeleton; male. Canines elongate.
 Andaman Islands. Zool. Society. 68, 3, 21, 7.

1497 *a*. Skull; adult male. Nose elongate, very slender; without
     lower jaw.
 Andaman Islands. 67, 6, 18, 1.   Presented by W. Theobald, Esq.

1497 *b*. Skull; adult female. Nose shorter, thicker. Wants canines.
     Occiput, upper part narrow.
 Andaman Islands. 67, 9, 28, 6.   Presented by W. Theobald, Esq.

1497 *e*. Skull; female. Wants one grinder.
 Andaman Islands. 70, 8, 17, 2.   Presented by Dr. Day.

1497 *f*. Lower jaw, discoloured. Sent with 1497 *a*.
 Andaman Islands.   Presented by W. Theobald, Esq.

8. SUS SCROFA, *Gray, Cat. Carniv. &c.* p. 337.

37 *a*. Animal, stuffed; half-grown, black.
 Europe.

37 *b*. Animal, stuffed; young.
 France. Domestic.

37 *c*. Animal; young.
 France, wild. Le Fevre. 43, 12, 29, 12.

37 *d*. Adult male.
 Europe.

37 *e*. Animal, stuffed.
 Darmstadt. 47, 3, 27, 39.

37 *f*. Animal, stuffed; just born.
 Europe. 60, 7, 22, 4.

37 *g* & *h*. Animal, unstuffed; just born.
 Germany. Born in Zool. Gardens. 58, 5, 26, 3 & 4.

713 *a*. Skin, unstuffed.
713 *a*. Skeleton of "*a*."
 "Sus scrofa barbarus," *Sclater, P. Z. S.* 1860, p. 43. *Gray, Cat.* 338.
 Barbary.

713 *w*. Animal, stuffed.
"Wild pig from Morocco," *Zool. Soc.*
Morocco. Zool. Soc. Presented by E. W. A. Drummond, Esq.
Skull taken out; lost at Zool. Soc.

713 *u*. Animal, stuffed; young female.
713 *u*. Skeleton and skull of "*u*." Changing teeth.
Barbary. 63, 12, 4, 1. Presented by H. Christy, Esq.

713 *c*. Skeleton.
Europe. Zool. Society.

713 *d*. Skeleton.
713 *m*. Skeleton; male.
Germany. Zool. Society.

713 *e*. Skull; young.
713 *n*. Skull.
Germany. Zool. Society.

713 *f*. Skull.
Europe. Zool. Society.

713 *j*. Skull.
Germany. Günther.

713 *u*. Skull.
Germany.

713 *k* & *i*. Two skulls of young.
Europe. Zool. Society.

713 *g*. Skull; nearly adult.
713 *l*. Skull, imperfect.
Europe.

9. Sus LIBYCUS, *Gray, Cat. Carniv. &c.* p. 338.

713 *a*. Skull; female.
Xanthus. Presented by Sir C. Fellows.

5. PORCULA, *Gray, Cat. Carniv. &c.* p. 339.

Face without any warts; ears ovate, covered with short hair.

1. PORCULA SALVIANIA, *Gray, Cat. Carniv. &c.* p. 340. *Hodgson, Journ. Asiatic Soc. Bengal,* xvi. t. xii., xiii., xvii., xxix.

The skull of *Porcula salviania*, from Mr. Hodgson, is not quite adult, the hinder grinder of each jaw being visible, but not developed. It has six grinders on each side in each jaw, and all the other characters of a *Sus*. Always without a tail.

1077 *c*. Animal, stuffed; young.
India, Saul Forest. Presented by B. H. Hodgson, Esq.

1077 *a*. Skull; half-grown female? and bones of feet.
India, Saul Forest. 54, 6, 3, 7. Presented by B. H. Hodgson, Esq.

1077 *b*. Skull (imperfect).
India. Presented by B. H. Hodgson, Esq.

K

1077 *d*. Skin, unstuffed.
1077 *d*. Skull of "*d*."
　India, Saul Forest. 58, 6, 24, 71. Presented by B. H. Hodgson, Esq.
1077 *e*. Skin, unstuffed.
1077 *e*. Skull of "*e*."
　Saul Forest. 58, 6, 24, 72. Presented by B. H. Hodgson, **Esq.**

　2. Domestic Swine. Ears more or less dependent, often very large.

6. SCROFA, *Gray, Cat. Carniv. &c.* p. 345.

　1. SCROFA DOMESTICA, *Gray, Cat. Carniv. &c.* p. 345.

713 *v*. Skull.
　Algiers. Christy. Zool. Society.

713 *b*. Skull.
　England.

717 *a*. Skull; old male.
　Sus gambianus, *Gray, List Mam. B. M.*
　Gambia.

713 *h*. Skull; old male.
　Africa.

713 *r*. Skull; old male.
　Hab. —?

713 *o* & *t*. Two nearly adult skulls.
　Hab. —? Jeude.

713 *p*. Skull; adult (diseased).
　Hab. —? Zool. Society.

716 *r*. Skull; young.
　India. Zool. Society.

713 *q*. Skull; young sow.
　Berkshire.

713 *e*. Skull; young.
　Domestic.

713 *s*. Skull; very young.
　Domestic.

713 *t*. Skull; adult.
　Holland. Domestic. Mus. Utrecht.

713 *w*. Skull.
　Holland. Domestic. Jeude.

7. CENTURIOSUS, *Gray, Cat. Carniv. &c.* p. 347.

　1. CENTURIOSUS PLICICEPS, *Gray, Cat. Carniv. &c.* p. 347; *P. Z. S.*
　　1863, p. 14, f. (skull and palate).
　Japanese Mask pig, *Bartlett, P. Z. S.* 1861, p. 263, f. (animal).

1387 *d*. Animal, stuffed; very young.
　Domestic. Zool. Soc. 62, 6, 28, 3.

1387 *a.* Skeleton ; very old animal.
Domestic. Japan. Bartlett.

1387 *b.* Skull ; young.
Domestic.

1387 *c.* Skeleton ; adult male.
Domestic. Zool. Gardens.

Tribe 3. BABIRUSSINA, *Gray, Cat. Carniv. &c.* p. 348; *Ann. &
Mag. Nat. Hist.* 1873.

Ears rounded, not pencilled at the end. Cutting-teeth $\frac{4-4}{6-6}$, the
front grinders separated from the upper canines by a long
diastema. The upper and lower canines of the male much elongated and recurved. The sheath of the upper canines elongate,
arising from the outer side of the margin of the upper jaw,
closely applied to but separated from the side of the nose, without
any or only a very slight indication of a cross ridge. Canines
not developed in the females, and their usual situation indicated
by a sharp-edged ridge just above the lower margin of the upper
jaw in front of the grinders. The males have a deep cavity on
each side of the roof of the hinder upper part of the inner
nostrils : this concavity is absent in the females, where this part
is only slightly concave. I cannot find any exit from these pits,
which are very deep. The bullæ of the ears are oblong and
elongate. No such concavity in the back of the nasal cavity of
any of the pigs that I have examined, but there is a deep pit on
each side of the centre of the hinder part of the nasal cavity in
the *Phacochœrus,* which is small in the young and larger in the
more adult skulls. In the adult skulls there is a very deep
concavity on each side of the roof of the inner nostrils, in front
of these pits which are separated from each other by a thin erect
longitudinal plate. These concavities are scarcely perceptible in
the skulls of the very young animals. The bullæ of the ears of
the skulls of the very young *Phacochœrus* are large, nearly
hemispherical, and very prominent, but in the adult skulls they
are small and scarcely separated from the rest of the skull.

8. BABIRUSSA, *Gray, Cat. Carniv. &c.* p. 349; *Ann. & Mag. Nat.
Hist.* 1873, p.

Intermaxillary bones moderate, scarcely produced behind. Canines
of the males elongate, convex at the sides. The lower ones rounded,
scarcely keeled in front, in the females wanting in the upper jaw, only
short conical and slightly recurved in the lower jaw.

1. BABIRUSSA ALFURUS, *Gray, Cat. Carniv. &c.* p. 349.
Sus babirussa, *Blainv. Ostéog.* t. ii. (skeleton ♀) & v. (skull and
teeth).

718 *v.* Animal, stuffed ; male.
718 *v.* Skeleton of " *v.* "
Celebes. Zool. Soc. 71, 5, 19, 7.

The adult skulls present two varieties in the canines, but I do not find any other character to separate them.

I. The upper and lower canines very long and gradually arched.

718 *i*. Skeleton (imperfect); male. Skull partly cut away. Canines small. Upper grinders much worn.
Malacca? Zool. Society.

718 *a*. Skull; male. Wanting one canine.
Malacca.

718 *b*. Skull; male, with part of skin.
Malacca. 38, 4, 16, 32.

718 *c*. Skull, with skin.
List Mam. p. 185.
Malacca.   Presented by Gen. Hardwicke.

718 *d*. Skull; male. Canines very large.
Malacca. 46, 3, 13, 3.

718 *e*. Skull; male, imperfect.
Malacca.

718 *f*. Skull; male, without canines, imperfect.
Hab. — ?

718 *g*. Skull; male. Canines large.
Malacca.   Presented by — Daniels, Esq.

718 *h*. Skull; male. Wanting canines and cutting-teeth.
Malacca. Zool. Society.

718 *j*. Skull.
Malacca. Zool. Society.

718 *k*. Skull; male. Canines large.
Borneo. Wallace. 59, 8, 16, 4.

718 *q*. Skull; male. Canines large, perfect.
Malacca. 67, 4, 12, 223. Jeude.

718 *p*. Skull; male.
Malacca.

II. The upper and lower canines shorter, not more than three inches long, the upper ones being very much curved, sometimes nearly in a circle.

718 *l*. Skull; adult. Canines very small.
Borneo. Wallace.

718 *m*. Skull; male. Canines small. Hinder palatal cavity small.
Malacca. Jeude. 60, 8, 27, 7.
Females, upper canines not developed.

718 *o*. Skull; female.
Malacca. Jeude. 67, 4, 12, 209.

Section 2. EURODONTINA.

Premolars deciduous, their places being filled up by the development of the molars; molars formed of laminæ, many rooted.

Family 1. PHACOCHŒRIDÆ, *Gray, Cat. Carniv. &c.* p. 352. *Ann. & Mag. Nat. Hist.* 1873.

1. PHACOCHŒRUS, *Gray, Cat. Carniv. &c.* p. 352.

Zygomatic arch very broad, with only a very slight broad concavity in front of the orbit. Lower canines triangular. The upper canines bent upwards and outwards, very large and thick, with a ridge across their sheath as in the *Suidæ*, but in both sexes. Lower canines flat on the outer sides and keeled in front. The sheath of the upper canines with a very obscure ridge across the middle in skulls said to belong to the two sexes, which were living in the Zoological Gardens. The sheath and upper canines of the female are rather smaller and more slender than those of the male. The upper and lower canines of the skulls with the milk-teeth even in the youngest, where the grinders are scarcely developed, have the canines well developed, slender and spreading out, the upper ones bent downwards and slightly reflexed at the end, the lower ones being recurved and most spread. The canines of the permanent series appear by the figure of the young female given by Mr. Sclater to be at first small, conical, but in this animal still in the Zoological Gardens (April, 1873) they have now become large and elongate, very similar to those of the males, but not quite so large as in that sex. Indeed the adult skulls received from the Zoological Gardens and from Mr. Sundevall, with the sex marked, show that the two sexes, unlike the pigs, have well-developed large canines. Both the skulls of the young animals above referred to, with their milk-teeth, have two well-developed cutting-teeth in the upper, and six in the lower jaw.

1. Phacochœrus æthiopicus, *Gray, Cat. Carniv. &c.* p. 353.
Phacochœrus? *Gray, Ann. & Mag. Nat. Hist.* 1870, vi. p. 190.

19 *a.* Animal, stuffed.
Phacochœrus Æliani, *Rüppell, Zool. Atlas*, t. xxv., xxvi.'
Abyssinia. Rüppell.

70 *c.* Animal, stuffed; adult male.
S. Africa.   Presented by the Earl of Derby.

74 *b.* Animal, stuffed; young.
S. Africa.

765 *b.* Animal, stuffed; adult female.
765 *b.* Skeleton of "*b.*" Cutting-teeth two above, fallen out.
P. Æliani, *Gerrard, Cat. Bones*, p. 280.
Africa.   62, 1, 22, 4.

765 *a.* Skin, unstuffed; adult female.
765 *a.* Skull of "*a.*"
"Phacochœrus Æliani, ♀," *Sundevall.*
Caffraria. Sundevall.   46, 6, 2, 74 & 75.

719 m. Skeleton; "female." Upper tusks rather thick.
P. æthiopicus, *Gerrard, Cat. Bones*, p. 280.
Africa. Zool. Society.

719 n. Skeleton; male. Upper tusks very large, thick and worn. One
cutting-tooth above.
P. æthiopicus, *Gerrard, Cat. Bones*, p. 280.
Africa. Zool. Society.

765 c. Skeleton.
P. Æliani, *Gerrard, Cat. Bones*, p. 280. *Jesse, P. Z. S.*
Abyssinia. Jesse.

719 a. Skull.
S. Africa.   Presented by J. C. Taunton, Esq.

719 b. Skull; very large; male. Tusks very thick.
Cape de Verd.   Presented by T. Tatum, Esq.

719 c. Skull; adult male. Back cut off.
Africa. 51, 11, 22, 1.

719 f, g, i, j, m. Skulls; without lower jaws.
S. Africa. Argent.

719 r. Skull; adult. Tusk very thick.
W. Africa. Baikic. 65, 5, 3, 5.

719 d. Skull, without incisors in upper jaw, and only form indications
of them in the lower.
S. Africa. Argent. 50, 8, 24.

719 e, h, k, l. Skulls, without lower jaws.
S. Africa. Argent. 50, 8, 24.

719 q. Skull; adult male. Tusks very large.
Abyssinia, near Zaulla. Blandford. 69, 10, 24, 47.

74 a. Skull; male, from skin in very bad state.
S. Africa.   Presented by W. Burchell, LL.D.

719 o. Skull; very young, with only three grinders above and below.
S. Africa. 71, 8, 3, 4.

719 p. Skull; very young, with only indications of the grinders.
P. æthiopicus, *Gray, Ann. & Mag. Nat. Hist.* 1871, p. 138.
S. Africa. 71, 7, 3, 5.

The young skulls are scarcely to be distinguished from those of the
genus *Sus* by their dentition, as the grinders are not worn and the
large permanent grinder is not developed, but are known by the dilata-
tion and spreading out of the hinder part of the base of the lower jaw.
The younger, which is 4¼ in. long, has only the second deciduous
grinder developed in the upper jaw and the first and second in the lower
jaw. The canines are slender and conical, curved downwards and outwards. The pulp of the two upper cutting-teeth is visible, but they are
not cut. The canines of the lower jaw slender, and the outer cutting-
teeth are alone visible.

The larger skull, which is 6½ in. long, has the small conical first, and
the second and third larger deciduous molars well developed, as are

also the two upper cutting-teeth, and the canines are like those of the smaller skull bent down, but the alveolar part of the base rather more produced. The lower jaw has the three deciduous grinders and the six cutting-teeth all well developed, the two middle ones being much the longest. The canines are, as in the smaller skull, slender and curved, the lower jaw is much more developed, extended in front, and broader and much more expanded below, approximating it more closely to the shape of the jaw of the adult animal. (See *Ann. & Mag. Nat. Hist.* 1871, viii. p. 183).

Sub-Section 1. HYDROTHERIUM.

Front part of jaws dilated and truncated. Nostrils on the upper side of nose closed by a valve. Eyes high up on the sides of the head in a line with the base of the ears.

Sub-Order 6. OBESA, *Gray, Cat. Carniv. &c.* p. 356.

Family 8. HIPPOPOTAMIDÆ, *Gray, Cat. Carniv. &c.* p. 356.

1. HIPPOPOTAMUS, *Gray, Cat. Carniv. &c.* p. 356.

1. HIPPOPOTAMUS AMPHIBIUS, *Gray, Cat. Carniv. &c.* p. 356.
Blainv. Ostéog. t. i.—vii.

726 *a.* Animal, badly stuffed.
S. Africa.   Presented by Mus. Roy. Coll. Surgeons.

726 *e.* Animal, stuffed; half-grown.
S. Africa.   42, 12, 6, 11.   Presented by the Earl of Derby.

726 *b.* Animal, stuffed; very young.
726 *d.* Animal, stuffed; young.
S. Africa.   42, 12, 6, 12.   Presented by the Earl of Derby.

726 *c.* Fœtus in spirit.
H. amphibius, *Gray, P. Z. S.* 1868, p. 491, t. 2; *Cat.* p. 357, f. 40
  (nat. size).
Cape of Good Hope.

726 *j.* Skeleton; young, imperfect.
Africa.

726 *a.* Skull; adult.
S. Africa.

726 *d.* Skull.
Gambia.   Presented by the Earl of Derby.

726 *f.* Skull.
S. Africa.   51, 11, 10, 12.   Zool. Society.

726 *g.* Skull; very young.   Length of upper jaw 28 in.
S. Africa.   51, 12, 23, 4.

726 *h.* Skull; very large.
Cape of Good Hope.   64, 11, 12, 51.   Said to be from the Linnean
  collection.

726 *i.* Skull.
Africa.   68, 2, 12, 1.   Presented by Dr. Falconer.

726 *k*. Skull.
  Natal.   69, 8, 13, 2.

726 *c*. Lower canine, very much elongated, subspiral.
  S. Africa.   46, 3, 19, 1.

726 *e*. Tusk.
  S. Africa.   41, 1, 13, 45.   Mantell.

726 *b*. Front of lower jaw.
  S. Africa.

2. CHÆROPSIS, *Gray, Cat. Carniv. &c.* p. 357.

  1. CHÆROPSIS LIBERIENSIS, *Gray, Cat. Carniv. &c.* p. 357.
  Young Liberian Hippopotamus, *Graphic*, 29th March, 1873.
  Hippopotamus liberiensis, *Milne-Edwards*, t. i., ii. (animal and
    skull).

1312 *a*, *b*. Cast of skull.
  W. Africa.   Liberia.   Presented by G. S. Morton, Esq.

  Section 2. HETEROGNATHA, *Gray, Cat. Carniv. &c.* p. 358.

  Front of both jaws contracted and bent down. The upper and
    lower cutting-teeth when present produced into the form of
    tusks. Teats pectoral. Limbs well developed, or rudimentary
    and fin-shaped.

  Sub-Order 7. PROBOSCIDEA, *Gray, Cat.* p. 358.

  Family 9. ELEPHANTIDÆ, *Gray, Cat. Carniv. &c.* p. 358.

1. ELEPHAS, *Gray, Cat. Carniv. &c.* p. 358.

  1. ELEPHAS INDICUS, *Gray, Cat. Carniv. &c.* p. 358.

707 *a*. Fœtus, in spirit.
  Seba, i. t. iii. f. 1.
  Gray, P. Z. S. 1868, p. 491, f. 1.   *Cat. Bruta*, p. 359, f. 41.
  Ceylon.   Presented by Capt. T. Surplen.

707 *b*. Animal, three parts grown, stuffed.
  India.   Zoological Society.

707 *c*. Skeleton; mounted.
  India.   Presented by Sir Jasper Nichols and Gen. Hardwicke.

707 *j*. Skeleton; disarticulated.
  India.   Zool. Society.

707 *h*. Skeleton; half-grown.
  India.   51, 11, 10, 16.

707 *a*. Skull of the variety called " Dauntelah."
  Corse, Phil. Trans. 1799, ii. p, 205.
  India.   Presented by J. Corse Scott, Esq.

707 *b*. Skull of the variety called " Mooknah."
  Corse, Phil. Trans. 1799, t. ii. p. 205.
  India.   Presented by J. Corse Scott, Esq.

707 *g*. Skull, without lower jaw; very young.
India. 47, 3, 5, 38.

707 *f*. Skull; half-grown, with tusks.
India. 47, 12, 10.

g. Skull.
India.

707 *h*. Skull, cut in two.
India. 47, 3, 5, 38.

707 *i*. Skull.
India, Meerat. 51, 9, 28, 1.   Presented by Lieut. G. Campbell.

707 *k*. Skull; very young.
India.

707 *l, m*. Teeth.
India.

707 *n, o*. Teeth.   Upper molar.
Ceylon.   Presented by Dr. J. Davy.

707 *p—s*. Teeth.
India. 41, 12, 12, 3.   Presented by H. Lacon, Esq.

707 *t, u*. Teeth.
India. 37, 6, 10, 262.   48, 12, 6, 7.

707 *v—z*. Teeth, in various stages of growth.
India.

707 *d*. Tusks.
Ceylon.   Presented by H.R.H. the Duke of Sussex.

707 *e*. The base of two tusks.
India.

707 *a²*. Tooth; lower molar, right side.
India. 37, 6, 10, 261.

707 *b²*. Tooth.
India. 49, 7, 27, 5.   Presented by J. E. Boileau, Esq.

707 *c²*. Section of an upper molar.
India. 57, 2, 14, 3.   Presented by J. E. Boileau, Esq.

707 *d², e²*. Sections of teeth.
India. 49, 7, 27, 6—7.   Presented by J. E. J. Boileau, Esq.

707 *f²*. Femur of young.
India.   Presented by R. Weeks, Esq.

707 *g², h²*. Upper and lower molar teeth.
India. 62, 7, 18, 10—11.

707 *i*. Lower molar teeth.
India.

1445 *a**. Skeleton.
Elephas sumatranus, *Temminck*.
1445 *b*. Skin from the above.
Sumatra, Padang. Mus. Leyden. 65, 1, 29, 1.

L

2. LOXODONTA, *Gray, Cat. Carniv. &c.* p. 359.

1. LOXODONTA AFRICANA, *Gray, Cat. Carniv. &c.* p. 359.
E. africanus, *Blainv. Ostéog.* t. iii., vii. & ix. (skull and teeth).

708 *a.* Animal, stuffed ; very young.
708 *b.* Skull of " *a.*"
Elephas africanus, *Falconer, Faun. Sevalensis,* t. 44.
S. Africa. 43, 12, 7, 31. Verreaux.

708 *h.* Skeleton.
S. Africa. Stevens. 58, 11, 15, 1.

708 *o.* Skull, without lower jaw.
Africa. 68, 2, 12, 2.   Presented by Dr. Faulkner.

708 *c.* Skull; half-grown. Wants lower jaw.
Africa. 51, 1, 4, 1. Williams.

708 *i.* Skull ; young.
Africa. Jamrach.

708 *j.* Skull of a fœtus.
Africa. 59, 12, 29, 5.   Presented by Dr. Livingstone.

708 *n.* Skull of fœtus.
E. Africa.   Presented by Dr. Kirk.

708 *a.* Tooth.
S. Africa.   Presented by John Lee, LL.D.

708 *d.* Two tusks.
Africa. 51, 1, 4, 23. Williams.

708 *e—g.* Teeth.
Africa. 54, 3, 28, 3—5.

708 *k.* Tusk (deformed).
Mozambique. 59, 2, 26, 1.   Presented by S. O. Soarez, Esq.

708 *l.* Tusk (deformed).
Africa. 64, 1, 25, 3. Livingstone.

708 *m.* Tusk (deformed).
Africa. 65, 3, 30, 10. Dr. Baikie's collection.

708 *p.* Four molar teeth.
S. Africa. 59, 12, 29, 3.   Presented by Dr. Livingstone.

708 *q.* Base of skull, with molar teeth and lower jaw.
S. Africa. 59, 12, 29, 1.   Presented by Dr. Livingstone.

708 *r.* Base of skull, with molar teeth and lower jaw.
S. Africa. 59, 12, 29, 2.   Presented by Dr. Livingstone.

Sub-Order 8. SIRENIA, *Gray, Cat. Seals and Whales,* 1865, p. 356.

Family 10. MANATIDÆ, *Gray, Cat. Seals and Whales,* p. 356.

Tribe 1. MANATINA, *Gray, Cat. Seals and Whales,* p. 357.

1. MANATUS, *Gray, Cat. Seals and Whales,* p. 357.

Vrolik, Dierkunde, iii. p. 53, t. i., ii., iii.

Pelvic bones very small, fins with three rudimentary hoof-like nails.
Eyes large, circular, partly hidden by the projecting front margin.

Iris circular, radiating pupil circular, half the diameter of the eye. Nose much produced on the sides, and truncated in front. Lower lip narrow, square, separated from the chin by an arched groove, and the chin from the throat by a circular fold.

1. MANATUS AUSTRALIS, *Gray, Cat. Seals and Whales*, p. 359.
M. americanus, *Vrolik, Dierkunde*, iv. 1852, t. i., vi.

370 *a*. Fœtus in spirits.
Jamaica. Mus. Sloane.

370 *b*. Animal, stuffed.
Surinam. Krauss. 64, 6, 5, 1.

370 *c*. Animal, stuffed ; very young.
Central America. Hulse.

370 *d*. Skin ♂ , belonging to the skeleton " 370 *f*."
Surinam. Krauss.

370 *e*. Animal, young.
S. America. Janson. 68, 2, 19, 8.

370 *e*. Skeleton ♀ .
Surinam. Krauss. 64, 6, 5, 2.

370 *f*. Skeleton ♂ .
Surinam. Krauss. 70, 8, 16, 4.

370 *e*. Skeleton.
'Manatu de Surinam,' *Krauss, Müller's Archiv. für Anat.* 1867,
p. 415, f. 30.
Surinam. Krauss.

370 *c*. Skeleton, imperfect.
Cuba. 57, 2, 6, 1.   Presented by H. Christie, Esq.

370 *b*. Skull.
Jamaica. Gosse. 47, 2, 2, 1.

370 *a*. Skull.
W. Indies. 43, 3, 10, 12.

370 *d*. Skull.
W. Indies. 56, 11, 30, 6.

2. MANATUS SENEGALENSIS, *Gray, Cat. Seals and Whales*, p. 360 ;
Ann. & Mag. Nat. Hist. 1865, xv. p. 134.
M. Vogelii, *Baikie, P. Z. S.* 1857, p. 38, t. 51 (skull).

1388 *a*. Animal, stuffed.
W. Coast of Africa.   Presented by Messrs. Forster.

1388 *b*. Stuffed.
Africa.

1388 *c*. Stuffed.
W. Africa.

1388 *a*. Skeleton.
Africa. 61, 7, 29, 22.

1388 *b*. Skeleton.
Africa. 61, 7, 29, 24.

1388 *c*. Skeleton, mounted.
W. Africa. 61, 7, 29, 23. Du Chaillu.

1388 *e*. Skeleton.
W. Africa. Du Chaillu, 1864. 64, 12, 1, 8.

1388 *f*. Skull.
Lagos. 65, 5, 8, 1.   Presented by W. MacCosky, Esq.

1388 *d*. Skull.
Manatus Vogellii, *Owen, Report Brit. Assoc.*
W. Africa. Baikie. 43, 3, 10, 12.   From the Royal Institution,
Liverpool.

1388 *g*. Skull.
W. Africa. Baikie. 65, 3, 30, 3.

2. HALICORE, *Gray, Cat. Seals and Whales*, p. 360.

1. HALICORE DUGONG, *Gray, l. c. f.* 361.

1027 *a*. Animal, stuffed; adult.
E. Indies?

1027 *b*. Animal, stuffed; adult.
E. Indies?

1027 *c*. Animal, stuffed; very young.
Malacca. 32, 5, 24, 1.

1027 *b*. Skeleton, skull wanting.
E. Indies? 55, 12, 22, 59.

1027 *c*. Skeleton, imperfect. Zool. Soc. Mus. Col.
Sumatra. Raffles. Zool. Soc. 55, 12, 22, 246.

1027 *a*. Skull; adult. 371 *a*.
India? Sumatra.   Presented by W. Elliot, Esq.

1027 *a*. Skeleton, imperfect.
Australia. Strange. 52, 6, 27, 1.

1027 *h*. Skeleton.
Moreton Bay.   Presented by J. Harris, Esq.

1027 *d*. Skull, wanting lower jaw.
Halicore australis, *Owen, Jukes, Voy.* H.M.S. 'Fly,' p. 225, f. 1, t. 27,
f. 3, 328, f. 5.
N.E. Coast Australia. 46, 8, 7, 19. Presented by J. B. Jukes, Esq.

1027 *b*, *c*. Two skulls, imperfect.
H. australis, *Owen, Jukes, Voy.* ' Fly,' p. 225, f. 1, t. 27, f. 3, 328, f. 5.
N.E. Coast Australia. 45, 7, 5, 25. 46, 7, 7, 19. Presented by J. B.
Jukes, Esq.

1027 *i*. Skull.
Moreton Bay. 63, 4, 10, 1.   Presented by H. B. Grundy, Esq.

1027 *f*. Skull.
Moreton Bay. 48, 8, 29, 6.   Presented by Capt. Stanley.

1027 *e*. Skull.
  Darnley Island, Torres Straits. Presented by the Earl of Derby.

1027 *g*. Skull.
  Moreton Bay. 48, 8, 29, 7. Presented by Capt. Stanley.

1027 *j*. Skull; young.
  Australia. 55, 3, 11, 1.

1027 *k*. Skull, imperfect.
  Australia. 48, 8, 29, 1.

2. HALICORE TABERNACULI, *Gray, Cat. Seals and Whales*, p. 364.

1534 *a*. Animal, stuffed.
  H. Tabernaculi, *Rüppell, Mus. Senckenb.* i. 113, t. 6.
  Red Sea, Coast of Abyssinia.

1534 *b*. Skin, stuffed.
1534 *b*. Skeleton of the above.
  Red Sea. Edward Gerrard, jun. 70, 8, 16, 1.

Tribe 2. RYTININA, *Gray, Cat. Seals and Whales*, p. 365.

3. RYTINA, *Gray, Cat. Seals and Whales*, p. 365.

1. RYTINA GIGAS, *Gray, Cat. Seals and Whales*, p. 365.

1394 *a* & *b*. Two ribs.
  Behring's Straits. From Mus. Petersburg.

INDEX

TO THE GENERA AND SPECIES OF PACHYDERMATA.

keitloa	... 51	pliciceps	... 66	sumatranus (C.)	47
labiatus	... 54	Porcula	... 65	sumatranus (R.)	35
laurillardi	... 34	porcus	... 56	stenocephalus ...	47
leucogenys	... 34	Potamochœrus ...	56	Sus	... 60
leucomystax	... 61	quagga	... 37	tabernaculi	... 77
liberiensis	... 72	Rhinaster	... 51	taivanus	... 61
libycus	... 65	Rhinoceros	... 45	Tapirus	... 33
Loxodonta	... 74	Rhinochœrus	... 35	terrestris	... 33
Manatus	... 74	rufescens	... 41	timorensis	... 60
mystaceus	... 62	Rytina	... 77	torquatus	... 53
nasalis	... 47	salviania	... 65	unicornis	... 46
niger	... 48	Scrofa	... 66	verrucosus	... 59
Notophorus	... 53	scrofa	... 64	vittatus	... 58
onager	... 37	semicircularis ...	44	vulgaris	... 36
oswellii	... 52	senegalensis	... 75	zebra	... 38
papuensis	... 60	simum	... 52		
Phacochœrus	... 69	siniaticus	... 42		

Order PECORA.

Gray, Cat. Rum. Mam. p. 1.

Sub-Order 1. CAVICORNIA, *Gray, Cat. Rum.* Mam. p. 6.

Section 1. LEVICORNIA, *Gray, Cat. Rum. Mam.* p. 6.

Family 1. BOVIDÆ, *Gray, Cat. Rum. Mam.* p. 7.

1. BOS, *Gray, Cat. Rum. Mam.* p. 7.

1. Bos TAURUS, *Gray, Cat. Rum.* Mam. p. 7; *Cat. Mam. B. Mus., Ungulata,* t. i. fig. 1 (skull).

83 *g.* White Scotch bull, stuffed.
Chillingham Park.   Presented by the Earl of Tankeville.

83 *h.* Polish bull, stuffed.
Vienna.

83 *i.* Hungarian bull, stuffed.
Vienna.

603 *h.* Bones of body.
Nepal.   Presented by B. H. Hodgson, Esq.

603 *i.* Skeleton; very bad, mounted.
Hab. —?   47, 10, 23, 1.

603 *r.* Skeleton. "Piedmontese bull."
Piedmont.   Zool. Society.

603 *q.* Fœtal skeleton.
Hab. —?   62, 12, 29, 11.

603 *c.* Skull; young; lower jaw wanting.
Hab. —?

603 *f, g.* Skulls of domesticated cattle of Nepal; male and female.
Nepal.   45, 1, 8, 96—7.   Presented by B. H. Hodgson, Esq.

603 *j.* Skull.
England.

603 *k.* Skull.
"Bœuf sans cornes," *Cuv., Oss. Foss.* iv. tab. 9, figs. 3, 4.
England.

603 *l.* Skull. "Sikkim domestic cattle."
India?   Presented by B. H. Hodgson, Esq.

603 *m.* Skull. "Lowland domestic cattle."
Sikkim.   Presented by B. H. Hodgson, Esq.

603 *s.* Skull, with horns.
Hab. —?   67, 4, 12, 281.

603 *t, u, v.* Skulls, with horns.
Hab. —?   Lidth de Jeude.   67, 4, 12, 282—5.

603 *w, x.* Skulls, without horns.
Hab. —?   Lidth de Jeude.   67, 4, 12, 284 and 290.

603 *n*, *o*. Horns.
  Hab. — ?   52, 9, 18, 17.   Zool. Society.

603 *a*. Horns, very thick and light.
  Central Africa.   Pres. by Capt. Clapperton and Major Denham.

603 *b*. Two separate horns, very large.
  S. Africa.

603 *c*. Horns on frontal bone, very long and slender.
  Hab. — ?

2. Bos chinensis, *Gray, Cat. Rum. Mam.* p. 8.

1601 *a*. Skull of a male.
  Bos Chinensis, *Swinhoe, P.Z.S.,* 1870, p. 648, fig. 6 (animal),
    figs. 7, 8 (skull).
  Formosa.   72, 2, 10, 67.   Swinhoe.

1601 *b*. Skull of heifer.
  Formosa.   72, 2, 10, 68.   Swinhoe.

3. Bos indicus, *Gray, Cat. Rum. Mam.* p. 9.

1067 *c*. Skeleton; imperfect.
  India.   54, 6, 8, 1.   Zool. Society.

1067 *d*. Skull; male.
  India.   55, 12, 26, 147.   Zool. Society.

1067 *e*. Skull.
  India.   55, 12, 26, 148.

1067 *f*. Skull of a calf.
  India.   58, 5, 4, 18.

4. Bos dante, *Gray, Cat. Rum. Mam.* p. 9.

1046 *d*. Skin.
1046 *c*. Skeleton of " *d*."
  Gambia.   53, 11, 31, 18.   From the Earl of Derby's collection.

1046 *a*, *b*. Skulls; male and female.
  Gambia.   48, 1, 13, 3—4.   Presented by the Earl of Derby.

2. BUBALUS, *Gray, Cat. Rum. Mam.* p. 9.

1. Bubalus buffelus, *Gray, Cat. Rum. Mam.* p. 9.

604 *a*. Animal; stuffed.
  Manilla.   Zool. Society.

604 *b*. Animal; stuffed; one day old.
  Born in the Zoological Gardens.   42, 10, 22, 1.

604 *b*. Skull, with horns.
  India.

604 *d*. Skull, with horns.
  India ?

604 *e*. Skull, with horns.
  India ?

604 *i*. Skull ; the horns polished.
  Neilgherries. Presented by Gen. Hardwicke.

604 *j*. Skull, with horns. " Domestic variety."
  India. 45, 1, 8, 143. Presented by B. H. Hodgson, Esq.

604 *k*, *l*. Skulls, with horns. " Wild."
  India. 45, 1, 8, 142—3. Presented by B. H. Hodgson, Esq.

604 *m*. Skull, with horns.
  India. 45, 1, 8, 147. Presented by B. H. Hodgson, Esq.

604 *o*. Skull, with horns. " Domestic variety."
  India. 45, 1, 8, 145. Presented by B. H. Hodgson, Esq.

604 *q*. Skull, with horns.
  Bos arnee, *Smith, Gray, P.Z.S.* 1855, p. 16, pl. xl.
  Assam. Presented by Col. J. Matthie.

604 *r*, *s*, *t*. Skulls, with horns.
  Hab. —? Lidth de Jeude. 67, 4, 2, 286—8.

604 *u*. Skull, with small horns. " Formosa buffalo."
  Formosa. 70, 2, 10, 66. Swinhoe.

604 *v*. Skull, with long thin horns.
  India? 67, 4, 12, 290. Lidth de Jeude.

604 *w*. Skull, with horns.
  India ? 68, 12, 20, 1.

604 *x*. Skull, with one horn wanting.
  India ?

604 *p*. Frontal bone, with horns.
  Sumatra. Sir Stamford Raffles.

604 *n*. Pair of horns ; very slender.
  India. 45, 1, 8, 146. Presented by B. H. Hodgson, Esq.

604 *a*. Horns; separate, thick, 48 inches long.
  India.

604 *c*. Horns; separate, 78 inches long.
  India.

604 *f*. Horns; separate, and slender.
  India.

604 *h*. Horns, on frontal bone ; young ; one horn wanting.
  India ?

  2. BUBALUS BRACHYCEROS, *Gray, Cat. Rum. Mam.* p. 10; *Cat. Ungulata, Brit. Mus.* t. i. fig. 2 (skull).

605 *c*. Skeleton ; young.
  Gambia. Presented by the Earl of Derby.

605 *a*, *b*. Skulls and horns.
  Central Africa. Pres. by Capt. Clapperton and Major Denham.

M

605 *d*. Skull, without horns and lower jaw.
    W. Africa ?   55, 5, 20, 4.   Presented by Dr. W. B. Baikie.
605 *e*, *f*. Skulls.
    W. Africa.   Baikie.

3. BUBALUS CENTRALIS, *Gray, Cat. Rum. Mam.* p. 11.

1555 *a*. Skull ; left horn and lower jaw wanting.
    W. Africa.   65, 3, 30, 1.   Baikie.

4. BUBALUS RECLINIS, *Gray, Cat. Rum. Mam.* p. 12.

1602 *a*. Horns, on frontal bone.
    Bubalus reclinis, *Gray, Cat. Ungulata, B. M.* t. ii. f. 3.
    Africa ?   Mus. R. Society.
1602 *b*. Skull, with horns.
    Africa.   72, 11, 11, 1.   Blyth.

5. BUBALUS CAFFER, *Gray, Cat. Rum. Mam.* p. 12 ; *Cat. Mam.
    Brit. Mus. Ungulata,* t. ii. f. 1, 2 (skull and horns).

606 *a*. Animal ; stuffed ; adult male.
    S. Africa.   Presented by the Earl of Derby.
606 *b*. Animal ; stuffed.   Calf.
    S. Africa ?
606 *c*. Animal ; skin ; young.
    S. Africa.   Verreaux.
606 *e*. Skeleton ; male.
    S. Africa.   50, 11, 22.   Zool. Society.
606 *b*. Skull, with horns.
    S. Africa.
606 *c*. Skull, with horns.
    S. Africa.   42, 4, 10, 14.
606 *d*. Skull, with horns.
    Africa ?   48, 7, 13.
606 *i*. Skull, with horns.
    Africa ?
606 *f*. Horns on frontal bone.
    S. Africa.   58, 5, 4, 9.   Zool. Society.
606 *g*. Horns on frontal bone.
    S. Africa.   52, 2, 15, 10.
606 *h*. Horns, half grown.
    Algoa Bay.   71, 7, 3, 8.

3. ANOA, *Gray, Cat. Rum. Mam.* p. 12.

    1. ANOA DEPRESSICORNIS, *Gray, Cat. Rum. Mam.* p. 13 ; *Cat.
        Ungulata, B. M.* t. iii. f. 1, 2 (skull, with horns).

607 *a*. Animal, stuffed; male.
  Celebes.

607 *b*. Animal, stuffed; female; in bad state.
  Celebes.

607 *h*. Skeleton; mounted.
  Celebes.

607 *a*. Horns on frontal bone.
  Celebes.   Presented by Gen. Hardwicke.

607 *b*. Skull, with horns.
  Celebes?

607 *e*. Skull; adult female.
  Celebes. Wallace.

607 *f*. Skull; young female.
  Mindanao. Wallace.

607 *c*. Horns on frontal bone.
  Celebes?

607 *d*. Horns on frontal bone.
  Celebes? Zool. Society.

607 *g*. Single horn; male.
  Celebes? 60, 8, 27, 10. Wallace.

4. BIBOS, *Gray, Cat. Rum. Mam.* p. 13.

  1. BIBOS FRONTALIS, *Gray, Cat. Rum. Mam.* p. 13; *Cat. Ungulata,
    Brit. Mus.* t. iii. f. 3 (skull and horn).

608 *a*. Animal, stuffed; young male.
  India? 66, 7, 7, 4. Zool. Society.

608 *d*. Skeleton; imperfect.
  Nepal.   Presented by B. H. Hodgson, Esq.

608 *e*. Skeleton of a young male.
  India? 66, 8, 7, 4. Zool. Society.

608 *f*. Skeleton.
  India? 68, 3, 21, 10. Zool. Society.

  2. BIBOS GAURUS, *Gray, Cat. Rum. Mam.* p. 13.

609 *a*. Animal, stuffed; adult, male.
  Nepal.   Presented by B. H. Hodgson, Esq.

609 *k*. Skeleton; imperfect.
  Nepal.   Presented by B. H. Hodgson, Esq.

609 *l, m*. Skeletons; imperfect.
  Nepal.   Presented by B. H. Hodgson, Esq.

609 *q*. Skeleton.
  E. Indies.   Presented by Capt. W. C. Robinson.

609 *a*. Skull, with horns; male.
  Nepal. 45, 1, 8, 99.   Presented by B. H. Hodgson, Esq.

609 *b*. Skull, with horns; female.
   Nepal. 45, 1, 8, 100.   Presented by B. H. Hodgson, Esq.

609 *c*. Skull, with horns.
   India. 37, 6, 10, 290.

609 *d*. Skull, with horns.
   India.   Presented by Gen. Hardwicke.

609 *n*. Skull, with horns; very large. "Old bull, shot on the
     Suhyadri Mountains, or Western Ghauts, 1850, by Capt.
     Wycliffe Thompson."
   India, W. coast. 58, 5, 4, 2.

609 *o*. Skull, with horns. "Young bull, shot on the Western Ghauts,
     1850, by Capt. Wycliffe Thompson."
   India, W. coast. 58, 5, 4, 3.

609 *p*. Skull, with horns. "Cow, young; shot on the Western Ghauts,
     1850, by Capt. Wycliffe Thompson."
   India, W. coast. 58, 5, 4, 4.

609 *r*. Skull, with horns; young.
   Tenasserim. 67, 5, 12, 3.   Presented by R. C. Beavan, Esq.

609 *s*. Skull, without horns.
   India?

609 *t*. Skull, without horns.
   India?

609 *e*. Horns; separate; polished.
   India?

609 *f*. Horns; separate.
   India?

609 *g*. Horns; separate; young.
   India?

609 *h, i*. Horns; separate.
   India?

609 *j*. Single horn.
   India?

   3. BIBOS BANTING, *Gray, Cat. Rum. Mam.* p. 13.

745 *a*. Animal, stuffed; male.
   Java. 46, 12, 15, 4. Franks.

745 *b*. Animal, stuffed; young.
   Java? Zool. Society.

745 *c*. Animal; skin; female.
   Java. 46, 12, 15, 5. Franks.

745 *a*. Skeleton; male.
   Java. 45, 1, 12, 5. Leyden Museum.

745 *b*. Skeleton; female.
   Java. 45, 1, 12, 7. Leyden Museum.

745 *c*. Skull.
   Java? 67, 4, 12, 613. Lidth de Jeude.

5. BISON, *Gray, Cat. Rum. Mam.* p. 14.

1. BISON BONASSUS, *Gray, Cat. Rum. Mam.* p. 14.

610 *a.* Animal, stuffed.
Lithuana.   Presented by the Emperor of Russia.

610 *b.* Animal, stuffed; half-grown female.
Lithuana.   57, 11, 30, 1.   Bartlett.

610 *c.* Skeleton; mounted.
Lithuana.   Presented by the Emperor of Russia.

2. BISON AMERICANUS, *Gray, Cat. Rum. Mam.* p. 14; *Cat. Ungulata,*
Brit. Mus. tab. iv. figs. 1, 2 (skull and horns).

851 *a.* Animal; male; stuffed.
N. America.

851 *b.* Animal; calf; stuffed.
N. America.   Zool. Society.

851 *c.* Animal; half-grown; stuffed.
N. America.   Zool. Society.

851 *d.* Animal; skin; half-grown.
N. America.   43, 11, 28, 1.   Pres. by the Hudson's Bay Company.

851 *a.* Skeleton.
N. America.   Presented by the Earl of Derby.

851 *b.* Skeleton; female.
N. America.   50, 11, 22, 102.   Zool. Society.

851 *c.* Skeleton; male.
N. America.   65, 12, 8, 24.   Zool. Society.

851 *d.* Skeleton; female.
N. America.   Zool. Society.

6. POËPHAGUS, *Gray, Cat. Rum. Mam.* p. 14.

1. POËPHAGUS GRUNNIENS, *Gray, Cat. Rum. Mam.* p. 14; *Cat.*
Ungulata, Brit. Mus. tab. iv. figs. 3, 4 (skull and horns).

611 *a.* Animal, stuffed; male.   Wild specimen.
Asia?

611 *b.* Animal, stuffed.   Domestic variety.
Asia?

611 *c.* Animal; skin; half-grown.   Domestic variety.
Asia.

611 *b.* Skeleton of a small domesticated specimen; very imperfect.
Nepal.   Presented by B. H. Hodgson, Esq.

611 *n.* Skeleton of a domesticated variety.
India.   67, 4, 22, 6.

611 *a.* Skull, with horns.
Asia.

611 *c.* Skull, with horns.
Asia.   45, 1, 8, 102.

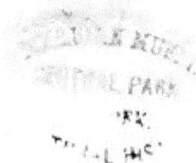

611 *d.* Skull, with horns.
   Asia. 45, 1, 8, 101.

611 *e.* Skull, with horns.
   Asia. 45, 1, 8, 101*.

611 *f.* Skull, with horns.
   Asia. 45, 1, 8, 103.

611 *g.* Horns on frontal bone.
   Asia. 45, 1, 8, 105.

611 *i.* Skull, with horns; young.
   Asia. 58, 6, 24, 152.

611 *o.* Skull. Left horn and lower jaw wanting.
   Asia. Transferred from the Palæontological Department.

611 *k.* Skull, with horns; female.
   Asia. 58, 6, 24, 152.

611 *h.* Horns on frontal bone.
   Asia.

611 *m.* Horn.
   Nepal. 58, 6, 24, 177.   Presented by B. H. Hodgson, Esq.

Family 2. ELEOTRAGIDÆ, *Gray, Cat. Rum. Mam.* p. 15.

1. KOBUS, *Gray, Cat. Rum. Mam.* p. 15.

   1. KOBUS ELLIPSIPRYMNUS, *Gray, Cat. Rum. Mam.* p. 15; *Cat. Ungulata, Brit. Mus.* tab. ix. figs. 3, 4 (skull and horns).

768 *a.* Animal, stuffed; adult male.
   S. Africa.

768 *b.* Animal, stuffed; young female.
   Africa.   Presented by the Earl of Derby.

768 *a.* Skull, with horns.
   Africa. 48, 3, 15, 1. Argent.

768 *b.* Horns on frontal bone.
   Africa. Warwick.

768 *c.* Skull; female; adult.
   Africa. 51, 12, 23, 7.

768 *d.* Skull; lower jaw wanting; female.
   Africa. 51, 12, 23, 8.

768 *e.* Skull; female.
   Africa. 55, 12, 26, 146. Zool. Society.

768 *f.* Skull, with horns; male.
   Africa. 59, 9, 23, 4.

768 *g.* Skull, with horns; male.
   Uzarano. 63, 7, 7, 9.   Presented by Capt. J. H. Speke.

768 *h.* Skull, with horns; female.
   Uzarano. 63, 7, 7, 10.   Presented by Capt. J. H. Speke.

2. KOBUS SING-SING, *Gray, Cat. Rum. Mam.* p. 15.

744 *a*. Animal; adult male; with very bad horns; stuffed.
Kobus sing-sing, *Knowsley Menagerie*, fig.
W. Africa. 61, 2, 10, 3. Zool. Society.

744 *b*. Animal; female; stuffed.
Gambia.   Presented by Earl of Derby.

744 *c*. Animal; young.
Gambia.   Presented by Earl of Derby.

744 *d*. Animal; young.
Gambia. 46, 10, 23, 4.   Presented by Earl of Derby.

744 *e*. Stuffed head; male.
Uganda. 63, 7, 7, 4.   Presented by Capt. Speke.

744 *a*. Skeleton; female.
Gambia.   Presented by Earl of Derby.

744 *d*. Skeleton; female; adult.
Gambia. Zool. Society.

744 *b*. Skull; young male; without horns.
Gambia. 46, 10, 17, 3.   Presented by Earl of Derby.

744 *c*. Skull; young female.
Gambia. 57, 2, 24, 4.   Presented by Earl of Derby.

3. KOBUS LEUCOTIS, *Gray, Cat. Rum. Mam.* p. 16.

1356 *c*. A stuffed head, with horns.
Central Africa. 60, 4, 20, 6. Petherick.

1356 *b*. A stuffed head, with horns.
Uganda. 63, 7, 7, 8.   Presented by Capt. J. H. Speke.

1356 *a*. Skull, with horns.
Central Africa. 60, 4, 20, 5. Petherick.

1356 *d*. Animal; skin; male; in very bad condition.
Awan, Bahr il Gazal. 59, 9, 23, 5.

4. KOBUS MARIA, *Gray, Cat. Rum. Mam.* p. 16.

a. A stuffed head, with horns.
Awan, Bahr il Gazal. 59, 9, 23, 8. Petherick.

b. A stuffed head; female.
Awan, Bahr il Gazal. 59, 9, 23, 9.

2. ADENOTA, *Gray, Cat. Rum. Mam.* p. 17.

1. ADENOTA KOB, *Gray, Cat. Rum. Mam.* p. 17.

885 *a*. Animal, stuffed; young male.
Gambia.   Presented by the Earl of Derby.

885 *b*. Animal, stuffed; female.
W. Africa.   Presented by Mr. E. Cross.

885 *c*. Animal, stuffed ; female, young.
Gambia.   Presented by the Earl of Derby.

885 *a*. Skull ; very young.
Gambia.   46, 11, 20, 9.   Presented by the Earl of Derby.

885 *b*. Skull, in a very imperfect state ; back of skull broken.
Gambia.   Presented by the Earl of Derby.

885 *c*. Skull, with horns ; male.
W. Africa.   65, 5, 3, 11.

885 *d*. Skull, with horns.
W. Africa.   63, 8, 4, 2.

3. ONOTRAGUS, *Gray, Cat. Rum. Mam.* p. 17.

　　1. ONOTRAGUS LECHEE, *Gray, Cat. Rum. Mam.* p. 17.

1357 *c*. Animal, stuffed ; male.
Banks of River Zouga.   Lat. 21°.   50, 7, 4, 2.
　　　　　　　　　　　　　　　Presented by Capt. Frank Vardon.

1357 *a*. Skull, with horn-cores only ; young.
Banks of the " Bahr il Gazal."   59, 0, 23, 6.   Petherick.

1357 *b*. Skull ; female.
Adenota lechee, female, *Gray, Cat. Rum. Mam.* t. ii. f. 4.
Banks of the " Bahr il Gazal."   59, 9, 23, 7.   Petherick.

4. ELEOTRAGUS, *Gray, Cat. Rum. Mam.* p. 18.

　　1. ELEOTRAGUS ARUNDINACEUS, *Gray, Cat. Rum. Mam.* p. 18.

630 *a*. Animal, stuffed ; male.
S. Africa.   46, 3, 23, 34.   Williams.

630 *b*. Animal, stuffed ; female.
630 *d*. Skull of " *b*."
S. Africa.   46, 5, 13, 19.

630 *c*. Skull.
S. Africa.   Presented by Dr. Burchell.

630 *h*. Skull, with horns ; male.
S. Africa.   51, 5, 5, 6.   Argent.

630 *i*. Skull ; female.
S. Africa.   Zool. Society.

630 *b*. Horns.
S. Africa.   46, 4, 2, 10.

630 *e*. Horns.
S. Africa.   Argent.

　　2. ELEOTRAGUS REDUNCUS, *Gray, Cat. Rum. Mam.* p. 18.

60 *a*. Animal, stuffed ; male.
Africa.   41, 12, 25, 3.

60 *b*. Animal, stuffed; female.
S. Africa.

60 *c*. Animal, stuffed; female.
S. Africa.

60 *d*. Animal, stuffed; young.
Gambia. 46, 10, 23, 17.   Presented by the Earl of Derby.

60 *d*. Skull, with horns; male.
S. Africa.   Presented by Dr. Burchell.

60 *f*. Skull, with horns; male.
S. Africa, Orange River. 51, 5, 5, 6.

60 *g*. Skull; female.
S. Africa, Orange River. 51, 5, 5, 7.

5. TETRACERUS, *Gray, Cat. Rum. Mam.* p. 18.

1. TETRACERUS QUADRICORNIS, *Gray, Cat. Rum. Mam.* p. 18.

628 *a*. Animal, stuffed; male.
Nepal. 43, 1, 12, 86.   Presented by B. H. Hodgson, Esq.

628 *b*. Animal, stuffed; male.
Nepal. 43, 1, 12, 87.   Presented by B. H. Hodgson, Esq.

628 *c*. Animal, stuffed; male.
India. 63, 5, 8, 4. Zool. Society.

628 *d*. Animal, stuffed; young female.
N. India. 69, 3, 5, 1. Zool. Society.

628 *h*. Skeleton?
India.

628 *a*. Horns on frontal bone.
N. India. 37, 6, 10, 68.

628 *b*. Horns on frontal bone.
India.

628 *c*. Skull; male; lower jaw wanting.
India. 45, 1, 8, 14.   Presented by B. H. Hodgson, Esq.

628 *d*. Skull; male; lower jaw wanting.
India. 38, 10, 29, 24.

628 *e*. Skull; male.
India. 56, 9, 22, 11.

628 *g*. Skull; male.
India. 58, 5, 26.   Presented by B. H. Hodgson, Esq.

2. TETRACERUS SUBQUADRICORNUTUS, *Gray, Cat. Rum. Mam.* p. 19.

43 *a*. Animal, stuffed; male.
Madras.   Presented by W. Elliot, Esq.

43 *b*. Animal, stuffed; male.
Madras. 46, 11, 6, 19.   Presented by T. C. Jerdon, Esq.

43 *c*. Animal, stuffed; female.
884 *a*. Skull of "43 *c*."
Madras. 46, 11, 6, 22.   Presented by T. C. Jerdon, Esq.

43 *d*. Animal, stuffed ; young female.
　Madras.　　　　　　　　　　Presented by W. Elliot, Esq

43 *e*. Animal, stuffed ; male.
　India.　55, 12, 24, 281.　Zool. Society.

6. CALOTRAGUS, *Gray, Cat. Rum. Mam.* p. 19.

　1. CALOTRAGUS MELANOTIS, *Gray, Cat. Rum. Mam.* p. 19.

46 *a*. Animal, stuffed ; male ; occiput brown.
　Cape of Good Hope.

46 *b*. Animal, stuffed ; female.
　Cape of Good Hope.　　　Presented by W. Burchell, Esq., LL.D.

46 *c*. Animal, stuffed ; female, young ; occiput with two black spots.
　S. Africa.

994 *b*. Skeleton of female.
　Africa.　Zool. Society.

994 *a*. Skull ; nasal bones wanting.
　S. Africa.　51, 5, 5, 18.

46 *d*. Animal, stuffed ; female ; whitish, base of some hairs reddish
　　　　grey ; two large diverging black stripes on occiput.

　C. melanotus pallida, *Gray, Cat. Ungul.* p. 72.
Cape of Good Hope.　Verreaux.

7. SCOPOPHORUS, *Gray, Cat. Rum. Mam.* p. 19.

　1. SCOPOPHORUS OUREBI, *Gray, Cat. Rum. Mam.* p. 19.

135 *a*. Animal, stuffed ; male ; one horn wanting.
　Cape of Good Hope.　S. African Museum.

135 *b*. Animal, stuffed ; female.
　Cape of Good Hope.　S. African Museum.

135 *c*. Animal, stuffed ; female.
775 *a*. Skull of " *c*."
　S. Africa.

135 *d*. Animal, stuffed ; female.
775 *b*. Skull of " *d*."
　S. Africa.　46, 1, 9, 13.

135 *e*. Animal, stuffed ; young female.
　S. Africa.　52, 12, 2, 6.

8. OREOTRAGUS, *Gray, Cat. Rum. Mam.* p. 20.

　1. OREOTRAGUS SALTATRIX, *Gray, Cat. Rum. Mam.* p. 20.

984 *b*. Animal, stuffed ; male.
　S. Africa.　44, 11, 8, 1.

984 *c*. Animal, stuffed ; male.
　S. Africa.

984 *d*. Animal, unstuffed.
　Abyssinia.　44, 5, 14, 37.

984 *e*. Animal, skin; male.
Abyssinia, Bogos country. 72, 2, 24, 14.

984 *f*. Animal, skin; female.
Abyssinia, Bogos country. 72, 2, 24, 15.

984 *a*. Skull.
Oreotragus saltatrix, *Gray, Cat. Mam. B. M.*, pt. iii. tab. 9, f. 2.
Cape of Good Hope.

9. CAPRICORNIS, *Gray, Cat. Rum. Mam.* p. 20.

1. CAPRICORNIS SUMATRENSIS, *Gray, Cat. Rum. Mam.* p. 20.

633 *c*. Animal, skin.
Manuhboom. 67, 5, 12, 1. Lieut. R. C. Beaven.

633 *d*. Animal, stuffed; half-grown.
Sumatra. 53, 8, 29, 55. Zool. Society.   Presented by Sir S. Raffles.

633 *a*. Frontal bone with horns.
Sumatra?

633 *b*. Skull.
Sumatra? 56, 5, 6, 78.   Presented by Prof. Oldham.

2. CAPRICORNIS BUBALINA, *Gray, Cat. Rum. Mam.* p. 20.

91 *b*. Animal, stuffed.
Nepal. 42, 4, 10, 14.

91 *c*. Animal, stuffed.
Nepal. 42, 4, 10, 15.

632 *c*. Skull, male.
C. bubalina, *Gray, Cat. Mam. Brit. Mus.* 1852, pt. iii. p. 111, tab. 13,
fig. 1.
Nepal.   Presented by B. H. Hodgson, Esq.

632 *d*. Skull, female.
Nepal. 46, 1, 8, 169.   Presented by B. H. Hodgson, Esq.

632 *f*. Skull.
Nepal. 55, 12, 26, 143.   Presented by B. H. Hodgson, Esq.

632 *g*. Skull.
Nepal. 58, 6, 24, 139.   Presented by B. H. Hodgson, Esq.

632 *h*. Skull; very imperfect.
Nepal.   Presented by B. H. Hodgson, Esq.

632 *i*. Skull, without horns.
India.   Presented by Dr. Falconer.

632 *j*. Skull, young.
India? 67, 4, 12, 224. Lidth de Jeude.

632 *k*. Animal, skin.
Nepal. 55, 1, 20, 2.
      Presented by His Highness Maharajah Dhuleep Singh.

632 *l*. Animal, skin.
Nepal. 55, 1, 20, 3
      Presented by His Highness Maharajah Dhuleep Singh.

632 *m*. Animal, skin.
  Nepal. 58, 6, 24, 21.   Presented by B. H. Hodgson, Esq.

  4. CAPRICORNIS SWINHOEI, *Gray, Cat. Rum. Mam.* p. 21.

1407 *a*. Animal, skin.
  Capr. Swinhoii, *Gray, P. Z. S.* 1862, p. 263, pl. xxxv
1407 *a*. Skull of " *a*."
  Formosa. 62, 12, 24, 2. Swinhoe.

1407 *b*. Animal, skin.
  Formosa. 70, 2, 10, 34. Swinhoe.

1407 *c*. Animal, skin.
  Formosa. 70, 2, 10, 35. Swinhoe.

1407 *b*. Skull.
  Formosa. 70, 2, 10, 33. Swinhoe.

  10. UROTRAGUS, *Gray, Cat. Rum. Mam.* p. 21.

    1. UROTRAGUS CAUDATUS, *Gray, Cat. Rum. Mam.* p. 21.

1562 *a*. Animal, stuffed.
1562 *a*. Skull of " *a*."
  Urotragus caudatus, *Gray, Cat. Rum. Mam.* tab. iv. fig. 8 (skull).
  N. China. 67, 12, 12, 2.

    Family 3. CEPHALOPHIDÆ, *Gray, Cat. Rum. Mam.* p. 21.

  1. GRIMMIA, *Gray, Cat. Rum. Mam.* p. 22.

    1. GRIMMIA NICTITANS, *Gray, Cat. Rum. Mam.* p. 22.

131 *b*. Animal, stuffed; male.
  S. Africa.

131 *a*. Animal, stuffed; female.
  S. Africa. 42, 4, 11, 2.

131 *c*. Animal, stuffed; young.
  S. Africa. 46, 61, 8.

131 *d*. Animal, stuffed; young.
  S. Africa. 46, 6, 1, 2.

626. *g—i*. Skeletons.
  S. Africa. Zool. Society.

626 *b*. Skull, without lower jaw.
  S. Africa.

626 *e*. Skull; male.
  S. Africa. 51, 5, 5, 11.

626 *f*. Skull; male, without horn-sheaths.
  S. Africa. 45, 11, 8, 36.

1366 *b*. Skull; female.
  S. Africa. 51, 5, 5, 13.

626 *c*. Skull; female.
  S. Africa. 51, 5, 5, 12.

626 *d*. Skull; female.
S. Africa. 46, 11, 18, 29.

2. GRIMMIA SPLENDIDULA, *Gray, Cat. Rum. Mam.* p. 22.

a. Animal, stuffed.
W. Africa, St. Paul de Loanda.   Presented by E. Gabriel, **Esq**.

3. GRIMMIA IRRORATA, *Gray, Cat. Rum. Mam.* p. 23.

1366 *a*. Animal, stuffed; male.
Port Natal. 46, 6, 2, 83. Stockholm Museum.

1366 *b*. Animal, stuffed; female.
Port Natal. 46, 6, 2, 82. Stockholm Museum.

1366 *c*. Skeleton.
S. Africa.

1366 *a*. Skull.
G. irrorata, *Gray, P. Z. S.* 1871, p. 591, fig. 1 (skull).
Port Natal. 46, 6, 2, 98. Stockholm Museum.

1366 *d*. Skull. 49, 1, 30, 29.
S. Africa.

4. GRIMMIA CAMPBELLIÆ, *Gray, Cat. Rum. Mam.* p. 23.

1336 *a*. Animal, stuffed; young.
Sierra Leone.   Presented by Mrs. A. G. Campbell.

5. GRIMMIA BURCHELLII, *Gray, Cat. Rum. Mam.* p. 24.

13 *a*. Animal, stuffed; male; without horn-sheath.
S. Africa.

13 *b*. Animal, stuffed; female.
S. Africa. 46, 6, 2, 81. Sundevall.

1358 *a*. Skull.
S. Africa.

1358 *c*. Animal, stuffed; young female.
S. Africa. 37, 9, 26, 120. Verreaux.

1358 *d*. Animal, stuffed; young.
S. Africa. 57, 6, 24, 2. Zool. Society.

6. GRIMMIA MADOQUA, *Gray, Cat. Rum. Mam.* p. 24.

1568 *a*. Animal, stuffed; male.
1568 *a*. Skull of "*a*."
Grimmia madoqua, *Gray, Cat. Rum. Mam.* tab. iv. f. 7.
Abyssinia. 71, 11, 29, 6.

2. TERPHONE, *Gray, Cat. Rum. Mam.* p. 24.

1. TERPHONE LONGICEPS, *Gray, Cat. Rum. Mam.* p. 24.

a. Skull, without horn-sheaths.
T. longiceps, *Gray, P. Z. S.* p. 204, fig. (skull).
W. Africa. 64, 12, 1, 12. Du Chaillu.

3. POTAMOTRAGUS, *Gray, Cat. Rum. Mam.* p. 25.

1. POTAMOTRAGUS MELANOPRYMNUS, *Gray, Cat. Rum. Mam.* p. 25.

1567 *a.* Animal, stuffed.
1567 *a.* Skull of " *a.* "
 Ceph. melanoprymnus, *Gray, P. Z. S.* 1871, pl. xliv. (animal), p. 594,
  f. 2, 3 (skull).
 W. Africa. 71, 5, 27, 1. Du Chaillu.

4. CEPHALOPHUS, *Gray, Cat. Rum. Mam.* p. 26.

1. CEPHALOPHUS SYLVICULTRIX, *Gray, Cat. Rum. Mam.* p. 26.

The concavity of the tear-bag deep, the deepest part placed near
the front of the orbit; varying in shape, but generally more or less
circular.

1139 *b.* Animal, stuffed; adult male.
 Sierra Leone. 55, 12, 24, 278. Zool. Society.

1139 *c.* Animal, stuffed; half-grown.
 Sierra Leone. 44, 9, 7, 2.   Presented by the Earl of Derby.

1139 *d.* Furrier's skin.
 W. Africa. 63, 6, 13, 7.

1139 *a.* Skull.
 C. sylvacultrix, *Gray, P. Z. S.* 1871, p. 596, fig. 4.
 Sierra Leone. 55, 12, 26, 161. Zool. Society.

2. CEPHALOPHUS BADIUS, *Gray, Cat. Rum. Mam.* p. 26.

Under side of the body rather paler bay; chest black; the under
side of the head paler whitish, the hinder part of the belly and inner
side of the thighs whitish.

418 *a.* Animal, stuffed.
 Sierra Leone. 46, 11, 4, 9.   Presented by the Earl of Derby.

418 *a.* Skull.
 Africa.

418 *b.* Animal, stuffed; female.
 Cephalophus breviceps, *Gray, P. Z. S.* 1866, p. 203, tab. xx.
 Africa. 69, 3, 5, 2. Zool. Soc. Coll.

418 *c.* Animal, skin; female. Dark bay Head, thighs and shoulders
  blackish. Dorsal streak, front and hind legs and tail above
  black. Tuft at end of tail pure white.
 Cameroons. 71, 7, 8, 7.

418 *b.* Skull; solid, with the broad concavity before the orbits large,
  deep, circular, separated from the margin of the jaw below,
  by a broad, convex, keeled space, separated by a defined line
  from the convex side of the nose. Intermaxillaries elongate

linear, scarcely broader above. **Nasals very much produced** behind between the frontals.

Africa.

3. **CEPHALOPHUS** AUREUS, *Gray, Ann. & Mag. N. Hist.* 1873, xii. p. 43.

Fur on the sides of the body rather elongate and soft; of the head and neck shorter and more ridged; pale bay. The crown of the head and long hairs between the horns; large oblong black spot over the eyes bright bay. Front part of the body and front part of the fore legs dark brown, varied with blackish hairs, which are more abundant, and form the indistinct streak down the back of the neck and front part of the back, but in the latter part they are more or less spread over the shoulders so as not to form any regular dorsal band. The black hairs are abundant on the front part of the fore legs; fore legs from the knees to the hoofs and the hind feet blackish. Under side of the head rather paler; chest blackish; hinder part of the belly and inner side of the thighs white. Tail dark brown, white on the under side; hoofs elongate.

Differs from *C. dorsalis* in the softness of the fur, the under side of the head being bay and not whitish; in the hind legs being brown and the hocks and feet to the upper hinder hoofs only being black; in many respects it is like *C. nigrifrons*, but the fur is softer and longer, and much paler.

1614 *a*. Animal, stuffed.

1614 *a*. Skull; solid, preorbital pit moderate, divided from the line of the grinders below by a moderate convexity, and rather deeper in front than the very compressed nose. The intermaxillary bones rather broader above. Nasals moderately long, truncated behind, and very broad, forming part of the orbital cavity, very acute in front.

C. aureus, Gray, Ann. & Mag. N. Hist. 1873, xii. p. 43.
Gaboon. Du Chaillu. 61, 7, 29, 11.

4. **CEPHALOPHUS RUFILATUS,** *Gray, Cat. Rum. Mam.* p. 26.

883 *a*. Animal, stuffed; male.
883 *a*. Skull of " *a*."
W. Africa. 62, 3, 19, 4. Zool. Society.

883 *b*. Animal, stuffed; female.
883 *c*. Skull of " *b*."
Cephalophus rufilatus, *Gray, P. Z. S.* 1871, p. 597, f. 5 (skull).
W. Africa. Cross.

883 *c*. Animal, stuffed; young.
883 *b*. Skull of " *c*."
Gambia. 46, 11, 2, 8.   Presented by the Earl of Derby.

883 *d*. Animal, skin.
Africa.

883 *e*. Skull; very young.
W. Africa. 65, 4, 27, 11. Baikie.

5. **CEPHALOPHUS DORSALIS**, *Gray, Cat. Rum. Mam.* p. 27.

1439 *a*. Animal, stuffed.
　Sierra Leone.　44, 11, 8, 13.

1439 *b*. Animal, skin.
　C. dorsalis, *Gray, P. Z. S.* 1871, pl. xlv. (animal).
1439 *a*. Skeleton of " *b*."
　Africa.　63, 12, 29, 1.　Zool. Society.

6. **CEPHALOPHUS NIGER**, *Gray, Cat. Rum. Mam.* p. 27.

a. Animal, stuffed.
　Coast of Guinea.　46, 2, 16, 2.　Franks.

7. **CEPHALOPHUS NATALENSIS**, *Gray, Cat. Rum. Mam.* p. 27.

132 *a*. Animal, stuffed ; male.
　Port Natal.　Kraus.

132 *b*. Animal, stuffed ; female.
　Port Natal.　Kraus.

132 *c*. Animal, stuffed ; young male.
　Port Natal.　46, 3, 23, 3.　Williams.

132 *d*. Animal, stuffed ; young.
　Port Natal.　Kraus.

901 *a*. Skull, with the left horn-sheath wanting.
　Cephalophus natalensis, *Gray, Cat. Mam. Brit. Mus.* iii. 1852, tab. x.
　　fig. 1.
　Port Natal.　46, 3, 26, 2.

8. **CEPHALOPHUS NIGRIFRONS**, *Gray, Cat. Rum. Mam.* p. 27.

1566 *a*. Animal, stuffed.
1566 *a*. Skull of " *a*."
　Cephalophus nigrifrons, *Gray, P. Z. S.* 1871, pl. xlvi. (animal),
　　p. 598, f. 6 (skull).
　Gaboon.　71, 5, 27, 2.　Du Chaillu.

9. **CEPHALOPHUS CORONATUS**, *Gray, Cat. Rum. Mam.* p. 28 ; *P. Z. S.*
　　1871, p. 599, f. 7 (skull).

624 *a*. Animal, stuffed.
624 *a*. Skull of " *a*."
　Gambia.　45, 10, 4, 3.　　　　　　Presented by the Earl of Derby.

624 *b*. Animal, stuffed.
624 *b*. Skull of " *b*."
　Gambia.　　　　　　　　　　　　Presented by the Earl of Derby.

624 *c*. Animal, stuffed.
　Gambia.　45, 10, 4, 2.　　　　　　Presented by the Earl of Derby.

624 *c*. Skull ; very young.
　W. Africa.　　　　　　　　　　　Presented by the Earl of Derby.

624 *d*. Skull ; nearly adult.
　W. Africa.　46, 11, 2, 24.　　　　Presented by the Earl of Derby.

10. CEPHALOPHUS WHITFIELDII, *Gray, Cat. Rum. Mam.* p. 28.

882 *a.* Animal, stuffed; young.
882 *a.* Skull of "*a.*"
    Gambia. 46, 11, 2, 6.   Presented by the Earl of Derby.

11. CEPHALOPHUS PYGMÆUS, *Gray, Cat. Rum. Mam.* p. 28.

1609 *a.* Animal, stuffed.
    S. Africa. 42, 4, 10, 8.

1609 *b.* Animal, stuffed; male.
    S. Africa.

1609 *c.* Animal, stuffed; young.
    S. Africa. 39, 9, 26, 122.

1609 *d.* Animal, stuffed; young.
    S. Africa. 69, 3, 5, 16. Verreaux.

1609 *e.* Animal, skin.
    S. E. Africa. 72, 10, 21, 5.

1609 *f.* Animal, skin.
    — Africa. 55, 12, 24, 394. Zool. Society.

1609 *a.* Skeleton.
    Africa. 60, 3, 8, 21. Zool. Society.

12. CEPHALOPHUS MAXWELLII, *Gray, Cat. Rum. Mam.* p. 28.

989 *a.* Animal, stuffed; male.
    W. Africa. 48, 10, 11, 5.   Presented by the Earl of Derby.

989 *b.* Animal, stuffed; female.
    W. Africa. 43, 12, 30, 10.   Presented by the Earl of Derby.

989 *c.* Animal, stuffed; male.
    W. Africa. 44, 2, 16, 3.

989 *d.* Animal, stuffed; two days old.
    Zool. Gardens. 71, 5, 19, 6. Zool. Society.

989 *f.* Animal, stuffed.
    Africa.

989 *g.* Animal, stuffed; female. Small size, with well developed
        conical horns.
    Zanzibar Island. 68, 9, 9, 1.   Presented by Dr. Kirk.

989 *e.* Skeleton.
    S. Africa. 60, 4, 18, 76.

989 *b.* Skull.
        C. Maxwellii, *Gray, P. Z. S.* 1871, p. 600, fig. 8.
    S. Africa. 61, 5, 5, 14.

989 *c.* Skull.
    S. Africa. 51, 5, 5, 15.

989 *d.* Skull.
    S. Africa. 55, 3, 11, 3.

13. **Cephalophus melanorheus**, *Gray, Cat. Rum. Mam.* p. 28.

1612 *a.* Animal, stuffed.
Fernando Po. 42, 11, 4, 30.   Pres. by Thomas Thomson, Esq.

1612 *b.* Animal, stuffed.
Fernando Po.   Presented by Thomas Thomson, Esq.

1612 *c.* Animal, skin; male.
Cameroon Mountains. 65, 3, 15, 8.   Presented by Capt. Burton.

1612 *d.* Animal, skin; male.
Cameroons. 62, 6, 28, 2.   Collected by Capt. Burton.

1612 *a.* Skull.
Cephalophus melanorheus, *Gray, Hand-List*, 1873, tab.

1612 *e.* Animal, skin; female; with horns.
Cameroon Mountains. 71, 7, 8, 6.

1612 *f.* Animal, skin; male.
Gaboon. 76, 5, 27, 4. Du Chaillu.

14. **Cephalophus punctulatus**, *Gray, Cat. Rum. Mam.* p. 29.

a. Animal, stuffed; young.
Sierra Leone.   Presented by Col. Sabine.

15. **Cephalophus bicolor**, *Gray, Cat. Rum. Mam.* p. 29.

1393 *a.* Animal, stuffed; young.
1393 *a.* Skull of "*a.*"
Natal, Umgozy Forest.   Presented by W. Fosbrook, Esq.

16. **Cephalophus Ogilbii**, *Gray, Cat. Rum. Mam.* p. 26.

The tear-bag oblong, elongated, with the deepest part in front, far away from the orbit. The deepest part of the concavity being in the lachrymal bone, which is large and with a large lobe in front. The nasal bones elongate, with a large lobe on each side over the lobe of the lachrymal bone.

Horns directed backwards, very rugose, on the inner front side of the base.

117 *a.* Animal, stuffed. The right fore foot wanting.
Fernando Po.   Presented by Thomas Thomson, Esq.

117 *b.* Furrier's skin.
Fernando Po. 55, 12, 24, 403. Zool. Society.

117 *c.* Animal, stuffed.
Gaboon. 61, 7, 29, 20. M. Du Chaillu.

1015 *a.* Skull; adult male.
W. Africa. 52, 2, 22, 10. L. Fraser.

117 *b.* Skull; young.
Gaboon. Du Chaillu.

117 *c.* Skull.
Africa. 64, 12, 21, 1.

117 *d.* Animal, stuffed; female.

>Cephal. leucogaster, *Gray, Ann. & Mag. N. Hist.* 1873, xii.
>p. 43.

117 *e.* Skull of "*d.*" Skull with a large circular tear-pit, which appears to be deeper in front of the centre and separated from the edge of the tooth-line by a broad convexity, rather keeled above. The intermaxillary bone broad for nearly its whole length, not quite reaching the nasal bones. The nasal bone with a broad rounded end which extends far up between the front ends of the frontal bones and **very acute** in front.

>Gaboon. Du Chaillu.

Family 4. PELEADÆ, *Gray, Cat. Rum. Mam.* p. 29.

1. PELEA, *Gray, Cat. Rum. Mam.* p. 29.

>1. PELEA CAPREOLA, *Gray, Cat. Rum. Mam.* p. 29, tab. iii. f. 5 (skull); *Cat. Mam. B. M.* pt. iii. tab. xxxvi. f. 4 (skull and horns).

44 *a.* Animal, stuffed; female.

>S. Africa.   Presented by W. Burchell, Esq., LL.D.

44 *b.* Animal, stuffed; female.

>S. Africa.

44 *c.* Animal, skin.

>S. Africa. 60, 7, 22, 1. Zool. Society.

629 *c.* Skeleton; female.

>S. Africa. 62, 12, 31, 10. Zool. Society.

629 *a.* Skull; male.

>S. Africa.   Presented by Dr. Burchell.

629 *b.* Skull; female.

>S. Africa.

629 *d.* Skull.

>Africa. 67, 4, 12, 222. Lidth de Jeude.

Family 5. NESOTRAGIDÆ, *Gray, Cat. Rum. Mam.* p. 30.

1. NANOTRAGUS, *Gray, Cat. Rum. Mam.* p. 30.

>1. NANOTRAGUS PERPUSILLUS, *Gray, Cat. Rum. Mam.* p. 30.   *V. Brooke, P. Z. S.* 1872, p. 637, t. 53, p. 642, fig.   (skull); *Cat. Mam. B. M.* pt. iii. (Ungulata) tab. x. f. 2 (head).

a. Animal, stuffed; young.

>W. Africa, Guinea. 44, 4, 4, 34.

b. Animal, in spirit.

>Fantee.   Presented by H. F. Blissett, Esq.

2. NESOTRAGUS, *Gray, Cat. Rum. Mam.* p. 30.

>1. NESOTRAGUS MOSCHATUS, *Gray, Cat. Rum. Mam.* p. 30.

1606 *a.* Animal, stuffed; male.

>Island of Zanzibar. 64, 3, 30, 1.   Presented by Capt. Speke.

1606 *b*. Animal, stuffed ; young male.
  Island of Zanzibar. 64, 3, 30, 2. Presented by Capt. Speke.

1606 *a*. Skull ; female.
  Zanzibar. 68, 3, 19, 1. Presented by Dr. Kirk.

1606 *b*. Skull ; female ; without nasal bones.
  Zanzibar. 68, 3, 19, 2. Presented by Dr. Kirk.

1606 *c*. Two horn-sheaths without the skull ; male.
  Zanzibar. Presented by Dr. Kirk.

2. NESOTRAGUS LIVINGSTONIANUS, *Gray, Cat. Rum. Mam.* p. 31.

1605 *a*. Skull.
  H. Livingstonianus, *Gray, Cat. Rum. Mam.* 1872, tab. 1, fig. 1.
  Central Africa. 68, 9, 28, 2. Presented by Dr. Kirk.

1605 *b*. Skull ; without lower jaw and one horn.
  Central Africa. 68, 9, 28, 3. Presented by Dr. Kirk.

1605 *c*. Frontal bone with horns and skin of face ; male.
  Central Africa. 64, 12, 19, 5. Presented by Dr. Kirk.

3. PEDIOTRAGUS, *Gray, Cat. Rum. Mam.* p. 31.

  1. PEDIOTRAGUS CAMPESTRIS, *Gray, Cat. Rum. Mam.* p. 31.

776 *a*. Animal, stuffed ; male.
  S. Africa. 46, 6, 2, 85. Stockholm Museum.

776 *b*. Animal, stuffed ; male.
776 *c*. Skull of " *b*."
  Port Natal. 45, 4, 5, 2. Warwick.

776 *c*. Animal, stuffed ; male.
  S. Africa. 46, 6, 1, 6.

776 *d*. Animal, stuffed ; female.
  S. Africa. 46, 6, 1, 7.

776 *e*. Animal, stuffed ; female.
776 *b*. Skull of " *e*."
  S. Africa. 46, 6, 2, 84.

776 *f*. Animal, stuffed ; female, young.
  Africa.

776 *g*. Animal, stuffed ; female, very young.
776 *e*. Skull of " *g*."
  S. Africa.

776 *h*. Animal, stuffed ; female, very young.
776 *a*. Skull of " *h*."
  S. Africa. 43, 12, 7, 24.

776 *d*. Skull ; male.
    P. campestris, *Gray, Cat. Mam. B. M., Ungulata,* tab. viii.
      f. 4, 5 (skull and horns).
  S. Africa. 46, 11, 18, 28.

776 *f*. Skull.
  E. Africa. Presented by Capt. Speke.

Family 6. OVIBOVIDÆ, *Gray, Cat. Rum. Mam.* p. 31.

1. OVIBOS, *Gray, Cat. Rum. Mam.* p. 31.

1. OVIBOS MOSCHATUS, *Gray, Cat. Rum. Mam.* p. 32; *Cat. Mam. B. M., Ungulata,* tab. v. figs. 1, 2 & 1—4 (skulls and horns).

612 *a*. Animal, adult, stuffed.
N. America. Parry. Presented by the Lords of the Admiralty.

612 *b*. Animal, half-grown, stuffed.
N. America.

612 *c*. Animal, skin; adult male (in a bad state).
N. America. 57, 3, 13, 2. Presented by Sir G. Simpson.

612 *d*. Animal, skin; without the skull.
N. America. Presented by Sir J. Richardson.

612 *c*. Skeleton; adult male; mounted.
N. America. J. Rae.

612 *d*. Skeleton (bones separate).
N. America. 55, 5, 14, 5. J. Rae. ·

612 *e*. Skeleton, young (bones separate).
N. America. 55, 5, 14, 7.

612 *f*. Skeleton; adult male (bones separate).
N. America. 55, 5, 14, 8. J. Rae.

612 *a, b*. Skulls, with horns.
N. America. Capt. Parry's expedition.
Presented by the Lords of the Admiralty.

612 *g, h*. Skulls, with horns; females.
N. America. 69, 2, 23, 1—2. Presented by A. G. Dallas, Esq.

2. BUDORCAS, *Gray, Cat. Rum. Mam.* p. 32.

1. BUDORCAS TAXICOLA, *Gray, Cat. Rum. Mam.* p. 32.

1603 *a*. Animal, stuffed.
Nepal. Presented by B. H. Hodgson, Esq.

1603 *b*. Skin.
Nepal. Presented by B. H. Hodgson, Esq.

1603 *a*. Horn on frontal bone.
India. 65, 6, 17, 1.

1603 *b, c*. Horns on frontal bone.
Assam.

1603 *d*. Horns on frontal bone.
B. taxicola, *Gray, Cat. Ung. B. M.* tab. v. fig. 3—4.
India.

Family 7. SAÏGADÆ, *Cat. Rum. Mam.* p. 32.

1. SAÏGA, *Gray, Cat. Rum. Mam.* p. 33.

  1. SAÏGA TARTARICA, *Gray, Cat. Rum. Mam.* p. 33; *Cat. Mam. B. M.,
Ungulata*, tab. vi. figs. 1, 2 (skull and horns).

96 *a.* Animal, stuffed; male.
  Siberia.　Brandt.

96 *b.* Animal, stuffed; young female.
  Russia.

96 *c.* Animal, skin; male (not in a good state).
  Russia.　51, 12, 28, 1.　Brandt.

96 *d.* Animal, skin; adult female.
  Russia.　73, 2, 24, 6.

96 *e.* Animal, skin; young.
  Russia.　65, 4, 28, 2.　Zool. Society.

613 *a.* Horns (separate).
  Russia.

613 *b.* Horns (separate).
  Russia.

613 *c.* Bones of the body only.
  Russia?　65, 4, 28, 2.　Zool. Society.

Family 8. PANTHOLOPIDÆ, *Gray, Cat. Rum. Mam.* p. 33.

1. PANTHOLOPS, *Gray, Cat. Rum. Mam.* p. 33.

  1. PANTHOLOPS HODGSONII, *Gray, Cat. Rum. Mam.* p. 33.

614 *a.* Animal, stuffed; male.
  Nepal.　　　　　　　　　　Presented by B. H. Hodgson, Esq.

614 *b.* Animal, skin; female.
  Nepal.　　　　　　　　　　Presented by B. H. Hodgson, Esq.

614 *a.* Horns (separate).
  Himalaya.　　　　　　　　Presented by B. H. Hodgson, Esq.

614 *c.* A single slender horn.
  Nepal.　　　　　　　　　　Presented by B. H. Hodgson, Esq.

614 *d.* Horns on frontal bone.
  India.

614 *e.* Skeleton, imperfect, without head or feet.
  India.　Abbott.

614 *b.* Skull of male.
     P. Hodgsonii, *Gray, Cat. Mam. B. M., Ungulata*, tab. vi.
     figs. 3, 4 (skull and horns).
  Nepal.　45, 1, 8, 136.　　　Presented by B. H. Hodgson, Esq.

Family 9. ANTILOPIDÆ, *Gray, Cat. Rum. Mam.* p. 34.

1. ÆGOCERUS, *Gray, Cat. Rum. Mam.* p. 34.

1. ÆGOCERUS LEUCOPHÆUS, *Gray, Cat. Rum. Mam.* p. 34; *Cat. Mam. B. M., Ungulata,* tab. xii. figs. 1, 2 (skull and horns).

636 *a.* Animal, stuffed; male.
S. Africa.   42, 4, 11, 9.   Smith.

636 *b.* Animal, stuffed; female.
S. Africa.   Presented by the Earl of Derby.

636 *a.* Horns.
S. Africa.

636 *c.* Horns on frontal bone.
S. Africa.

636 *d.* Horns.
S. Africa.

636 *g.* Horns, separate.
S. Africa.   50, 8, 24.

636 *b.* Skull with horns; female.
S. Africa.

2. ÆGOCERUS KOBA, *Gray, Cat. Rum. Mam.* p. 35.

1610 *a.* Horns.
Gambia.   Presented by the Earl of Derby.

1610 *b.* Horns.
Gambia.   Presented by the Earl of Derby.

3. ÆGOCERUS NIGER, *Gray, Cat. Rum. Mam.* p. 35.

1038 *a.* Animal, stuffed; male.
  Aigocerus niger, *Harris, Trans. Zool. Soc.* ii. 213, t. 39.
S. Africa.   Harris.

1038 *b.* Animal, stuffed; male.
S. Africa.

1038 *c.* Animal, stuffed; young male; red.
S. Africa.

1038 *d.* Animal, skin; young.
Africa.

1038 *c.* Horns; female.
S. Africa.   71, 7, 3, 9.

1038 *b.* Skeleton.
S. Africa.   67, 7, 8, 30.   Zool. Society.

1038 *a.* Skull.
S. Africa.   52, 9, 22, 1.   Gordon Cumming.

2. ORYX, *Gray, Cat. Rum. Mam.* p. 35.

1. ORYX GAZELLA, *Gray, Cat. Rum. Mam.* p. 35 ; *Cat. Mam. B. M.,*
Ungulata, tab. xii. figs. 3, 4 (skull and horns).

637 *a*. Animal, stuffed ; female (in a bad condition).
S. Africa.

637 *b*. Animal, stuffed ; male ; very young.
S. Africa.

637 *c*. Animal, stuffed ; just born.
S. Africa. 51, 7, 10, 26.

637 *d*. Animal, skin.
S. Africa. 46, 6, 1, 2.

637 *b, c.* Heads, stuffed, with horns.
S. Africa.

637 *a*. A single horn.
S. Africa.   Presented by Major Denham.

637 *e*. Horns.
S. Africa.

2. ORYX BEISA, *Gray, Cat. Rum. Mam.* p. 35.

a. Animal, skin.
Abyssinia. 71, 11, 29, 7.

b. Skull.
Abyssinia. 71, 11, 29, 7.

3. ORYX LEUCORYX, *Gray, Cat. Rum. Mam.* p. 36.

638 *a*. Animal, stuffed ; female.
Sennaar. 46, 6, 15, 62. L. Parreys.

638 *b*. Animal, stuffed ; very young.
638 *d*. Skull of " *b*."
Sennaar. 46, 6, 15, 63. L. Parreys.

638 *b*. Single horn.
Africa.

638 *e*. Skeleton.
Africa. Zool. Society.

638 *a*. Skull with horns.
N. Africa.

4. ORYX BEATRIX, *Gray, Cat. Rum. Mam.* p. 36.

1583 *a*. Animal, stuffed ; male.
1583 *a*. Skeleton of " *a*."
Oryx beatrix, *Gray, P. Z. S.* 1857, p. 157 tab. lv. (animal).
Bombay? Arabia?

1583 *b*. Animal, skin.
Bombay?? Arabia? 72, 11, 18, 1. Zool. Society.

3. ADDAX, *Gray, Cat. Rum. Mam.* p. 36.

    1. ADDAX NASOMACULATA, *Gray, Cat. Rum. Mam.* p. 36; *Cat. Mam. B. M., Ungulata,* tab. xii. fig. 5 (head and horns).

639 *a.* Animal, stuffed; female.
639 *b.* Bones of body of " *a.*"
    N. Africa.   Presented by the Earl of Derby.

639 *b.* Animal, stuffed; just born.
    Africa.   Presented by the Earl of Derby.

639 *a.* Horns (not united).
    N. Africa.   Presented by Capt. Clapperton and Major Denham.

639 *d.* Horns on frontal bone; young.
    Africa.

639 *c.* Skeleton; male.
    N. Africa.   Zool. Society.

4. PROCAPRA, *Gray, Cat. Rum. Mam.* p. 37.

    1. PROCAPRA GUTTUROSA, *Gray, Cat. Rum. Mam.* p. 37.

615 *a.* Animal, stuffed; male.
    Russia. 43, 12, 19, 1.

615 *b.* Animal, stuffed; female.
    Russia. 45, 4, 21, 7.

615 *c.* Animal, skin; young male.
    China. 70, 7, 18, 3. Swinhoe.

615 *d.* Animal, skin; male.
615 *c.* Skull of " *d.*"
    Pekin. 70, 2, 10, 37. Swinhoe.

615 *e, f.* Horns on frontal bone.
    Mongolia. 67, 1, 8, 3—4.   Presented by Dr. Lockhart.

615 *a.* Skull; female; from the stuffed specimen.
    Siberia. Brandt.

615 *b.* Skull; male.
    Pekin. 70, 2, 10, 95. Swinhoe.

615 *d.* Skull; male (imperfect).
    Pekin. 70, 2, 10, 96. Swinhoe.

    2. PROCAPRA PICTICAUDATA, *Gray, Cat. Rum. Mam.* p. 38.

781 *a.* Animal, stuffed; male.
    Thibet. 53, 8, 16, 18.   Presented by B. H. Hodgson, Esq.

781 *b.* Animal, stuffed; female.
    Thibet. 53, 8, 16, 19.   Presented by B. H. Hodgson, Esq.

781 *c.* Animal, skin; male.
    Thibet. 58, 6, 24, 10.   Presented by B. H. Hodgson, Esq.

781 *d.* Animal, skin.
    Thibet. 52, 12, 15, 6.

P

781 *a, b.* Skulls; male; without lower jaw; imperfect.
Thibet. 48, 6, 11, 19—20. Presented by B. H. Hodgson, Esq.

781 *c.* Skull; male; without lower jaw.
Thibet. 48, 6, 11, 21. Presented by B. H. Hodgson, Esq.

781 *d.* Skull; male; without lower jaw.
Thibet. Presented by B. H. Hodgson, Esq.

781 *e.* Skull; male.
Thibet. 52, 12, 15, 8.

781 *f.* Skull; male.
Asia. 56, 10, 1, 1.

781 *g.* Skull; female.
Asia. 56, 10, 1, 2.

5. GAZELLA, *Gray, Cat. Rum. Mam.* p. 38.

1. GAZELLA DORCAS, *Gray, Cat. Rum. Mam.* p. 38; *Cat. Mam. B. M.,*
Ungulata, tab. vi. fig. 5 (skull and horn).

129 *b.* Animal, stuffed; male.
Shores of the Red Sea? 38, 4, 16, 19.

129 *c.* Animal, stuffed; male.
Shores of the Red Sea.

129 *d.* Animal, stuffed; male.
Shores of the Red Sea. 44, 11, 28, 3. Pres. by the Earl of Derby.

129 *e.* Animal, stuffed.
Egypt. Presented by the Earl of Derby.

129 *f.* Animal, stuffed; female.
Shores of the Red Sea. Presented by Edw. Cross, Esq.

129 *g.* Animal, stuffed; young female.
Egypt. 46, 1, 19, 1. Presented by W. Tyler, Esq.

129 *h.* ? Animal, skin; female.
Ambor-Samhar. 200 ft. 69, 10, 24, 2.
Presented by W. T. Blanford, Esq.

616 *a—d.* Horns; male (separate).
Africa.

616 *h.* Horns; female.
Africa.

616 *j.* Horns; female.
Africa.

616 *n.* Skeleton.
Africa. 50, 11, 22, 34.

616 *o.* Skeleton.
Africa. 61, 8, 21, 1. Zool. Society.

616 *k.* Bones of the body.
Africa. 46, 5, 14, 42.

616 *q.* Skull; male; without lower jaw.
Abyssinia. 69, 10, 24, 101. Presented by W. T. Blanford, Esq.

616 *i*. Skull; female.

Egypt. Presented by Dr. Turnbull Christie.

616 *m*. Skull; female; imperfect; lower jaw and horn wanting.
Africa. 48, 2, 1, 28.

1*. GAZELLA CUVIERI, *Gray, Cat. Ung. B. M.* p. 57.

a. Animal, stuffed.
Gazella Cuvieri, *Ogilby, P. Z. S.* 1840, p. 34.
Mogador. 53, 8, 29, 47. Presented by the Zool. Society.

b. Animal, stuffed.
Africa. Presented by the Zool. Society.

c. Animal, stuffed.
Africa. 57, 3, 17, 1. Hultse.

2. GAZELLA ISABELLA, *Gray, Cat. Rum. Mam.* p. 38.

412 *a*. Animal, stuffed; male.
Kordofan.

412 *b*. Animal, stuffed; female.
412 *a*. Skull of " *b*."
"Antilope montana," *Parreys MSS*.
Kordofan. 46, 6, 15, 64. Parreys.

412 *c*. Animal, stuffed; female.
"Antilope montana," *Parreys MSS*.
Kordofan. 46, 6, 15, 64. Parreys.

412 *d*. Animal, stuffed; female.
Kordofan.

412 *e*. Animal, stuffed; young.
Kordofan? 59, 9, 12, 3. Zool. Society.

3. GAZELLA SUBGUTTUROSA, *Gray, Cat. Rum. Mam.* p. 38.

130 *a*. Animal, stuffed; female.
Siberia. Brandt.

130 *b*. Animal, stuffed; young female.
Siberia? 54, 12, 6, 4. Zool. Society.

4. GAZELLA SŒMMERINGII, *Gray, Cat. Rum. Mam.* p. 39.

1516 *a*. Animal, stuffed; male.
Abyssinia. Rüppell.

1516 *b*. Animal, stuffed; female.
Sennaar. 46, 6, 2, 79.

1516 *c*. Animal, skin; male.
Abyssinia, Bogos. 73, 2, 24, 7.

1516 *a*. Skull, with horns; male.
Abyssinia. Jesse.

5. GAZELLÁ MOHR, *Gray, Cat. Rum. Mam.* p. 39.

a. Animal, stuffed; male.
Antilope Mhorr, *Bennett, Trans. Z. Soc.* i. pl. 1, 1835.
Mogadore. 55, 12, 20, 279. Zool. Society.

6. GAZELLA RUFICOLLIS, *Gray, Cat. Rum. Mam.* p. 39.

a. Animal, stuffed; female.
Sennaar. 46, 6, 2, 78. Sundevall.

b. Animal, skin; male; not in a good state.
Kordofan. 48, 8, 18, 1. Parreys.

7. GAZELLA RUFIFRONS, *Gray, Cat. Rum. Mam.* p. 39.

411 *a.* Animal, stuffed; young male.
Africa.

411 *b.* Animal, stuffed; female.
411 *a.* Skull of " *b.*"
Senegal. 46, 11, 20, 8.

411 *c.* Animal, stuffed; female.
W. Africa.

411 *d.* Animal, stuffed; male.
Senegal. 46, 1, 10, 4.   Presented by the Earl of Derby.

411 *e.* Animal, stuffed; young.
Africa. 44, 1, 18, 24.

411 *f.* Animal, stuffed; young.
411 *a.* Skull of " *f.*"
Africa. 50, 2, 27, 16.   Presented by the Earl of Derby.

411 *g.* Animal, stuffed; adult male.
Abyssinia, Bogos. 73, 2, 24, 8.

411 *h, i.* Animals, skins; males.
Abyssinia, Bogos. 73, 2, 24, 9—10.

411 *j.* Animal, skin; female.
Abyssinia, Bogos. 73, 2, 24, 11.

6. TRAGOPS, *Gray, Cat. Rum. Mam.* p. 39.

1. TRAGOPS BENNETTII, *Gray, Cat. Rum. Mam.* p. 39.

617 *a.* Animal, stuffed; male.
Madras. 42, 8, 6, 9.   Presented by Col. Sykes.

617 *b.* Animal, stuffed; female.
Madras. 42, 8, 6, 10.   Presented by Col. Sykes.

617 *c.* Animal, stuffed; female.
India. 54, 12, 7, 1.   Presented by Miss Hearsey.

617 *a*. Horns of male.
India.   Presented by Dr. Turnbull Christie.

617 *b*. Skull, male; without the lower jaw.
N. India, Salt Range. 56, 5, 6, 69.   Presented by Prof. Oldham.

617 *c*. Skull; male.
N. India, Salt Range. 56, 5, 6, 70.   Presented by Prof. Oldham.

617 *d*. Skull; female; one horn wanting.
N. India, Salt Range. 56, 5, 6, 71.   Presented by Prof. Oldham.

617 *e, f*. Skulls; females; without lower jaws.
N. India, Salt Range. 56, 5, 6, 72—3.   Presented by Prof. Oldham.

7. ANTIDORCAS, *Gray, Cat. Rum. Mam.* p. 40.

1. ANTIDORCAS EUCHORE, *Gray, Cat. Rum. Mam.* p. 40.

39 *a*. Animal, stuffed; male.
S. Africa. S. African Museum.

39 *b*. Animal, stuffed; male.
S. Africa.   Presented by the Earl of Derby.

39 *c*. Animal, stuffed; young female.
S. Africa.

39 *d*. Animal, stuffed; female.
S. Africa. 46, 10, 26, 19. Turner.

618 *a, b*. Horns.
Africa.

618 *d*. Horns on frontal bone.
Africa.

618 *f*. Skeleton; young.
S. Africa. 53, 11, 30, 19. Zool. Society.

618 *g*. Skeleton; adult male.
S. Africa. 59, 2, 11, 3. Zool. Society.

618 *c*. Skull and horns.
S. Africa.   Presented by W. Burchell, Esq.

618 *e*. Skull.
Africa.

8. ANTILOPE, *Gray, Cat. Rum. Mam.* p. 40.

1. ANTILOPE BEZOARTICA, *Gray, Cat. Rum. Mam.* p. 40; *Cat. Mam.
B. M., Ungulata*, tab. viii. f. 1, 2, 3 (skull and horns).

97 *b*. Animal, stuffed; male.
India. 42, 9, 20, 1.

97 *g*. Animal, stuffed; male.
Madras. 38, 3, 13, 31.   Presented by Walter Elliot, Esq.

97 *a*. Animal, stuffed; female.
India. 51, 7, 3, 10. Cross.

620 *a*, *b*, *c*. Horns.
India.

620 *d*. Skull, with horns.
India.

620 *e*. Horns.
India.

620 *h*. Horns on frontal bone.
India.   45, 1, 8, 138.   Presented by B. H. Hodgson, Esq.

620 *j*. Horns on frontal bone.
India.   48, 7, 13, 11.

620 *k*. Horns on frontal bone.
India.

620 *n*. Horns on frontal bone.
India.   47, 7, 19, 16.

620 *r*. Skeleton ; young male.
India.

620 *g*. Skull, with horns.
India.   45, 1, 18, 139.   Presented by B. H. Hodgson, Esq.

620 *o*. Skull ; female.
India.   58, 5, 4, 600.   Zool. Society.

620 *p*. Skull ; female.
India.   64, 4, 22, 3.   Zool. Society.

9. NEOTRAGUS, *Gray, Cat. Rum. Mam.* p. 40.

1. Neotragus Saltiana, *Gray, Cat. Rum. Mam.* p. 40 ; *Cat. Mam.
B. M., Ungulata*, tab. ix. f. 3.

94 *a*. Animal, stuffed ; male.
Abyssinia.   55, 12, 24, 285.   Rüppell.   Zool. Society.

94 *b*. Animal, stuffed ; male.
Abyssinia.

94 *c*. Animal, stuffed ; female.
Abyssinia.

94 *d*. Animal, stuffed ; male.
Abyssinia.   61, 2, 30, 8.   Harris.

94 *e*. Animal, stuffed ; female.
Abyssinia.   61, 2, 30, 9.   Harris.

94 *g*. Animal, skin ; male.
Abyssinia, Anseba Valley, 4000 ft.   69, 10, 24, 4.   Blanford.

94 *h*. Animal, skin ; female.
Abyssinia, Anseba Valley, 4000 ft.   69, 10, 24, 3.   Blanford.

621 *b*. Skeleton ; female.
Abyssinia.   69, 2, 2, 10.   Jesse.

621 *a*. Skull; female.
Neotragus saltiana, *Gray, Cat. Rum. Mam.* tab. 1, f. 2.
Abyssinia.

10. NEMORHEDUS, *Gray, Cat. Rum. Mam.* p. 41.

1. **NEMORHEDUS GORAL**, *Gray, Cat. Rum. Mam.* p. 41; *Cat. Mam. B. M., Ungulata,* tab. xiii. f. 2.

24 *b*. Animal, stuffed; male.
Nepal.

24 *c*. Animal, stuffed; male.
Nepal.

24 *d*. Animal, stuffed; young.
Nepal. 37, 6, 10, 16.

24 *e*. Animal, skin; male.
Nepal. 45, 1, 8, 326.   Presented by B. H. Hodgson, Esq.

24 *f*. Animal, skin; male.
Nepal. 45, 1, 8, 327.   Presented by B. H. Hodgson, Esq.

634 *m*. Horns.
India.   Presented by Prof. Oldham.

634 *a*. Horns; female.
India.

634 *b*. Skull.
Nepal. 45, 1, 8, 183.   Presented by B. H. Hodgson, Esq.

634 *c*. Skull, without the horn-sheaths.
Nepal. 45, 1, 8, 181.   Presented by B. H. Hodgson, Esq.

634 *d*. Skull; female; the nasal bones wanting.
Nepal. 45, 1, 8, 182.   Presented by B. H. Hodgson, Esq.

634 *h*. Skull; without horn-sheaths.
Kashmir.   Presented by Prof. Oldham.

634 *j*. Skull; without horn-sheaths or lower jaw.
India. 56, 5, 6, 76. Zool. Society.

634 *k*. Skull; young.
India. 60, 4, 22, 5.

634 *l*. Skull; without horn-sheaths.
Nepal. 58, 5, 4, 34.   Presented by B. H. Hodgson, Esq.

11. MAZAMA, *Gray, Cat. Rum. Mam.* p. 41.

1. **MAZAMA AMERICANA**, *Gray, Cat. Rum. Mam.* p. 41; *Cat. Mam. B. M., Ungulata,* tab. xiv. f. 1 (head and horns).

a. Animal, stuffed.
Mazama americana, *Richardson, Fauna Bor.* p. 268.
Rocky Mountains. Zool. Society.

b. Animal, stuffed.
Rocky Mountains. Zool. Society.

c. Animal, skin.
   Summer Camp, Rocky Mountains.   63, 2, 24, 43.
                                        Presented by J. K. Lord, Esq.

d. Animal, skin ; without a skull.
   Fort Pilby.   52, 11, 30, 4.   Presented by J. Rae, Esq.

12. RUPICAPRA, *Gray, Cat. Rum. Mam.* p. 41.

   1. RUPICAPRA TRAGUS, *Gray, Cat. Rum. Mam.* p. 41 ; *Cat. Mam.
        B. M., Ungulata,* tab. xiv. f. 2, 3, 4 (skull and horns).

95 *a.* Animal, stuffed ; male ; in winter coat.
   Alps.

95 *b.* Animal, stuffed.
   Alps.

95 *c.* Animal, stuffed ; in summer coat.
   Alps.

95 *d.* Animal, stuffed ; young.
631 *k.* Skull of "*d.*"
   Alps.   51, 7, 10, 25.

631 *b.* Horns on frontal bone.
   Alps.   Presented by Gen. Hardwicke.

631 *h—j.* Horns.
   Alps.   46, 10, 13, 31—32—33.

631 *d, e.* Horns.
   Alps.

631 *g.* Skeleton.
   Alps.   Brandt.

631 *f.* Skull.
   Alps.   46, 7, 7, 2.   Presented by J. Gould, Esq.

631 *a.* Skull.
   Alps.   Presented by Gen. Hardwicke.

631 *l.* Skull, without horns ; male.
   Switzerland.   Günther.

631 *m.* Skull ; without lower jaw.
   Alps.

631 *n.* Skull ; one horn wanting.
   Alps.   47, 4, 12, 221.   Lidth de Jeude.

   Family 10. ÆPYCEROTIDÆ, *Gray, Cat. Rum. Mam.* p. 42.

1. ÆPYCEROS, *Gray, Cat. Rum. Mam.* p. 42.

   1. ÆPYCEROS MELAMPUS, *Gray, Cat. Rum. Mam.* p. 42 ; *Cat. Mam.
        B. M., Ungulata,* tab. vii. f. 3 (head and horn).

107 *a.* Animal, stuffed ; male.
   S. Africa.   Smut.

107 *b*. Animal, stuffed ; female.
S. Africa. Smut.

107 *c*. Animal, stuffed ; young male.
S. Africa. Smut.

107 *d*. Animal, stuffed ; female.
S. Africa.   Presented by the Earl of Derby.

619 *a*. Horns on frontal bone.
S. Africa.   Presented by Dr. Burchell.

619 *b, c*. Horns on frontal bone.
S. Africa. 48, 7, 13, 8—9. Warwick.

619 *f*. Horns on base of skull ; young.
S. Africa.   Presented by — Hora, Esq.

619 *g*. Skull, with horns ; adult ; without lower jaw.
S. Africa. 59, 8, 17, 1.

619 *h*. Skull ; young male.
E. Africa. 63, 7, 7, 13.   Presented by Capt. Speke.

Family 11. CONNOCHETIDÆ, *Gray, Cat. Rum. Mam.* p. 42.

1. CONNOCHETES, *Gray, Cat. Rum. Mam.* p. 43.

  1. CONNOCHETES GNU, *Gray, Cat. Rum. Mam.* p. 43 ; *Cat. Mam. B. M., Ungulata*, tab. xv. f. 45.

645 *a*. Animal, stuffed ; half-grown.
S. Africa.

645 *b*. Animal, stuffed ; young.
S. Africa. 46, 7, 2, 5.

645 *c*. A stuffed head ; young, with straight horns.
645 *h*. Skull of "*c*."
Africa. 69, 8, 11, 5.

645 *a*. Horns on frontal bone.
S. Africa.   Presented by J. Hillier, **Esq.**

645 *d*. Skeleton.
S. Africa. 50, 11, 22, 70. Zool. Society.

645 *b*. Skull.
S. Africa. 48, 8, 29, 1.

645 *c*. Skull.
S. Africa. 48, 8, 29, 2.

645 *e*. Skull, without lower jaw.
Africa.

645 *g*. Skull.
S. Africa. 59, 5, 6, 1.   Presented by T. Butler, Esq.

2. GORGON, *Gray, Cat. Rum. Mam.* p. 43.

　1. Gorgon fasciatus, *Gray, Cat. Rum. Mam.* p. 43.

138 *a.* Animal, stuffed; male.
　S. Africa.　S. African Museum.

138 *b.* Animal, stuffed; female.
　S. Africa.

138 *c.* Animal, stuffed; young.
　S. Africa.　S. African Museum.

767 *a.* Skull.
　S. Africa.　48, 7, 13, 1.

767 *b.* Skull; adult male.
　Uzarano.　63, 7, 7, 6.　　　　　　Presented by Capt. Speke.

767 *c.* Skull; very young.
　Uzarano.　63, 7, 7, 7.　　　　　　Presented by Capt. Speke.

　　Family 12. DAMALIDÆ, *Gray, Cat. Rum. Mam.* p. 43.

1. ALCELAPHUS, *Gray, Cat. Rum. Mam.* p. 43.

　1. Alcelaphus bubalis, *Gray, Cat. Rum. Mam.* p. 43.

641 *a.* Animal, stuffed; male.
　N. Africa.　Zool. Society.

641 *b.* Animal, stuffed; young.
　N. Africa.　45, 10, 30, 153.

641 *c.* Animal, skin; male.
　Abyssinia, Bogos.　73, 2, 24, 12.

641 *d.* Animal, skin; young.
　Abyssinia, Bogos.　73, 2, 24, 13.

641 *b.* Horns on frontal bone.
　N. Africa?

641 *c.* Skeleton.
　N. Africa.　59, 2, 10, 1.　Zool. Society.

641 *a.* Skull, with horns (one deformed).
　N. Africa?

641 *d.* Skull.
　Central Africa.　59, 9, 23, 2.　Petherick.

641 *e.* Skull.
　S. Africa.　46, 7, 2, 2.

　2. Alcelaphus major, *Gray, Cat. Rum. Mam.* p. 44.

a. Horns; male and female.
　Boselaphus major, *Blyth, P. Z. S.* 1869, p. 51, fig. A, 1 and 2.
　Africa.　69, 2, 9, 1—2.　Blyth.

3. ALCELAPHUS CAAMA, *Gray, Cat. Rum. Mam.* p. 44; *Cat. Mam.
   B. M., Ungulata,* tab. xvi. f. 1, 2, 3 (horns).

640 *a.* Animal, stuffed; male.
   S. Africa.   42, 4, 11, 6.   S. African Museum.

640 *b.* Animal, stuffed.
   S. Africa.   Presented by the Earl of Derby.

640 *c.* Animal, stuffed; young.
   S. Africa.   46, 7, 11, 4.   Warwick.

640 *a—e.* Horns.
   S. Africa.

640 *f, g.* Horns.
   S. Africa.   48, 7, 13, 3—4.   Warwick.

640 *l, m, o.* Horns.
   S. Africa.

640 *h.* Skull; lower jaw wanting.
   Africa.

640 *i.* Skull; imperfect.
   S. Africa.

640 *p.* Skull; male.
   S. Africa.

4. ALCELAPHUS LICHTENSTEINII, *Gray, Cat. Rum. Mam.* p. 44.

1355 *a.* Skull, imperfect; without the lower jaw.
   E. Africa.   60, 1, 10, 19.

1355 *b.* Skull, imperfect; without the horn-sheaths or lower jaw.
   E. Africa.   64, 7, 16, 1.   Dalton.

2. DAMALIS, *Gray, Cat. Rum. Mam.* p. 45.

   2. DAMALIS LUNATUS, *Gray, Cat. Rum. Mam.* p. 45.

642 *a.* Animal, stuffed; male.
   S. Africa.   42, 4, 11, 5.   S. African Museum.

642 *b.* Animal, stuffed; female.
   S. Africa.   S. African Museum.

642 *a.* Horns.
   S. Africa.   Presented by Dr. Burchell.

642 *b.* Horns.
   S. Africa.   48, 7, 13, 5.   Warwick.

642 *c.* Horns.
   S. Africa.   50, 8, 24, 1.

   2. DAMALIS SENEGALENSIS, *Gray, Cat. Rum. Mam.* p. 45; *Cat.
   Mam. B. M., Ungulata,* tab. xvi. figs. 4, 5 (skull and horns).

643 *a.* Skull.
   Bornou.   Presented by Capt. Clapperton and Major Denham.

643 *b*. Skull; lower jaw wanting.
Bornou.　　　　　Presented by Capt. Clapperton and Major Denham.

643 *c, d*, Skulls.
Gambia.　　　　　　　　　Presented by the Earl of Derby.

643 *e, f*. Skulls.
Bohr il Gazal.　Petherick.

643 *g*. Skull; very young.
Gambia.　　　　　　　　　Presented by the Earl of Derby.

　　3. DAMALIS PYGARGA, *Gray, Cat. Rum. Mam.* p. 45, tab. iii. fig. 6
　　　　(skull).

644 *a*. Animal, stuffed; male.
S. Africa.　Smut.

644 *b*. Animal, stuffed; female.
S. Africa.　Smut.

644 *c*. Animal, stuffed; young; face brown.
S. Africa.　39, 9, 26, 124.　Verreaux.

644 *d*. Animal, stuffed; young; face brown.
S. Africa.　43, 9, 27, 26.　Brandt.

644 *a*. Horns.
S. Africa.　　　　　　　　　Presented by Dr. Burchell.

644 *b*. Horns.
Africa.

644 *c, d*. Horns.
S. Africa.　Warwick.

644 *e*. Horns.
S. Africa.　48, 7, 19, 16.

644 *f*. Horns, long and slender.
Africa.

644 *h, i, k, l*. Horns.
S. Africa.

644 *q*. Skeleton.
Africa.　Zool. Society.

644 *r*. Skull.
S. Africa.

644 *s*. Skull; lower jaw wanting.
S. Africa.

644 *n*. Skull; male.
S. Africa.　58, 3, 17, 3.

644 *t*. Skull; female.
S. Africa.　58, 3, 17, 4.

　　4. DAMALIS ALBIFRONS, *Gray, Cat. Rum. Mam.* p. 45.

a. Animal, stuffed; female.
S. Africa.　Zool. Society.

5. **Damalis?** zebra, *Gray, Cat. Rum. Mam.* p. 45.

a. A furrier's skin, without head or limbs.
   Antelope? *Bennett, P. Z. S.* 1832, p. 122.
   A. Doria, *Ogilby, P. Z. S.* 1836, p. 121.
   Sierra Leone. 38, 4, 16, 327.   Presented by the Zool. Society.

b. A furrier's skin, without head or limbs.
   Sierra Leone. 55, 12, 24, 294.

Section II. ANGULICORNIA, *Gray, Cat. Rum. Mam.*, p. 46.

Family 13. STREPISICEROTIDÆ, *Gray, Cat. Rum. Mam.* p. 46.

1. STREPSICEROS, *Gray, Cat. Rum. Mam.* p. 46.

   1. Strepsiceros kudu, *Gray, Cat. Rum. Mam.* p. 46; *Cat. Mam.*
       B. M., Ungulata, tab. xvii. figs. 1, 2 (skull and horn).

646 *a.* Animal, stuffed; male.
   S. Africa.

646 *b.* Animal, stuffed; female.
646 *i.* Skull of "*b.*"
   S. Africa. 46, 6, 15, 31. Warwick.

646 *c.* Animal, stuffed; very young.
   S. Africa. 46, 6, 1, 4.

646 *h.* Skin.
   S. Africa. 61, 8, 21, 8. Zool. Society.

646 *d.* Skin; male.
   S. Africa.

646 *e.* Skin; female.
646 *i.* Skull of "*e.*"
   Abyssinia. 71, 11, 29, 1.

646 *f, g.* Skins; males.
   Abyssinia. 71, 11, 29, 2—3.

646 *n—o.* Skulls.
   Africa.

646 *q.* Horns, very wide apart at the tips.
   Africa. 71, 9, 26, 1.

646 *b.* Horns on frontal bone.
   S. Africa.

646 *d.* Horns on frontal bone.
   Africa.

646 *h.* Single horn.
   "Damalis canna," *Ham. Smith.*
   S. Africa.

646 *e, f.* Skulls; males.
   Africa.

646 *j.* Skull with horns.
   S. Africa. 48, 7, 11, 1.

646 *k*. Skull of a young male, without lower jaw.
S. Africa.

2. STREPSICEROS TENDAL, *Gray, Cat. Rum. Mam.* p. 46.

a. Animal, stuffed ; male.
Abyssinia.   61, 2, 30, 10.   Presented by the East India Company.

2. OREAS, *Gray, Cat. Rum. Mam.* p. 47.

1. OREAS CANNA, *Gray, Cat. Rum. Mam.* p. 47 ; *Cat. Mam. B. M.,
Ungulata*, tab. xvii. figs. 3, 4 (skull and horn).

647 *a*. Animal, stuffed ; male.
S. Africa.   Presented by the Earl of Derby.
647 *b*. Animal, stuffed ; female.
S. Africa.   Presented by the Earl of Derby.
647 *c*. Animal, stuffed ; young.
Africa.   43, 9, 27, 25.
647 *d*. Animal, stuffed ; young.
Africa.   63, 12, 3, 6.   Zool. Society.
647 *a—c*. Horns.
S. Africa.
647 *h*. Horns on frontal bone.
S. Africa.   Zool. Society.
647 *j*. Horns on frontal bone.
Algoa Bay.   61, 2, 3, 3.   Wemys.
647 *e*. Skeleton.
S. Africa.   Presented by the Earl of Derby.
647 *i*. Skeleton.
S. Africa.
647 *f*. Skull ; young.
S. Africa.   56, 7, 9, 1.
647 *g*. Skull ; very young.
S. Africa.

2. OREAS DERBIANUS, *Gray, Cat. Rum. Mam.* p. 47.

1648 *a*. A flat skin, without legs.
Senegal.   63, 4, 15, 1.   Presented by F. W. Reade, Esq.
1648 *a, b*. Horns.
Gambia.   Presented by the Earl of Derby.
1648 *c, d*. Skulls ; male and female.
Senegal.   Presented by F. W. Reade, Esq.

3. EURYCEROS, *Gray, Cat. Rum. Mam.* p. 47.

1. EURYCEROS EURYCEROS, *Gray, Cat. Rum. Mam.* p. 48.

852 *a.* Animal, stuffed; young male (without hoofs).
　　Tragelaphus albovirgatus, *Du Chaillu.*
　Ashkankoloo Mountains.　M. Du Chaillu.

852 *c.* Skull, imperfect.
　Africa.

852 *a.* Horns.
　　Antilope eurycerus, *Ogilby, P. Z. S.* 1836, p. 120.
　W. Africa?　Zool. Society.

852 *b.* Imperfect skull, with horns.
　Africa.　Zool. Society Mus. Coll.

2. EURYCEROS ANGASII, *Gray, Cat. Rum. Mam.* p. 48.

1170 *a.* Animal, stuffed; male.
　N. side of the Pongi River.　58, 3, 4, 1.　Eastwood.

1170 *b.* Animal, stuffed; male.
　Zulu Country, near St. Lucia Bay.　　Pres. by R. S. Fellowes, Esq.

1170 *c.* Animal, stuffed; female.
　Zulu Country, near St. Lucia Bay.　　Pres. by R. S. Fellowes, Esq.

1170 *d.* Animal, skin; male; in a bad state.
　　"Antilopus roualeynii," *Gordon.*
　Zambesi.　60, 2, 11, 14.　Chapman.

1170 *e.* Animal, skin; female.
　N. side of the Pongi River.　58, 3, 4, 2.　Eastwood.

1170 *f.* Animal, skin; young female.
　　"Myala antelope," *Proudfoot.*
　Africa.　50, 8, 30, 2.　Proudfoot.

1170 *b.* Horns, separate.
　Gaboon.　52, 2, 26, 30.　Parzudaki.

1170 *a.* Skull; male.
　Africa.　50, 8, 30, 1.　Proudfoot.

3. EURYCEROS SPEKII, *Gray, Cat. Rum. Mam.* p. 48.

a. Animal, stuffed; male; very young.
　Karayweh, E. Africa.　63, 7, 7, 2.　　Pres. by Capt. J. H. Speke.

b. Animal, skin; male; imperfect.
　Africa.　65, 5, 9, 20.　　Presented by M. Du Chaillu.

a. Horns.
　Karayweh, E. Africa.　63, 7, 7, 1.　　Presented by Capt. Speke.

b. Horns.
　Bight of Biafra.　　Presented by Lieut. Allen.

c. Horns.
　Lak Gnami.　　Presented by J. A. Green, Esq.

d. Single horn.
  Africa.   Parzudaki.

4. TRAGELAPHUS, *Gray, Cat. Rum. Mam.* p. 50.

  1. Tragelaphus scripta, *Gray, Cat. Rum. Mam.* p. 50; *Cat. Mam.
      B. M., Ungulata,* tab. xviii. figs. 1, 2 (skull and horn).

413 *a.* Animal, stuffed; male.
  W. Africa.

413 *b.* Animal, stuffed; young male.
  W. Africa.

713 *c.* Animal, stuffed; young.
  W. Africa.   46, 2, 28, 1.   Presented by the Earl of Derby.

413 *c.* Skeleton; male.
  W. Africa.   52, 11, 5, 3.

413 *e.* Skeleton; male.
  W. Africa.   61, 1, 18, 8.   Zool. Society.

413 *f.* Skeleton; imperfect.
  W. Africa.   63, 5, 3, 10.   Dalton.

413 *a.* Skull, with horns; adult.
  W. Africa, Gambia.   46, 11, 2, 22.   Pres. by the Earl of Derby.

413 *b.* Skull, without horn-sheaths.
  W. Africa, Gambia.   46, 11, 2, 23.   Pres. by the Earl of Derby.

413 *d.* Skull; young.
  Senegal.   45, 10, 4, 1.   Presented by the Earl of Derby.

413 *g, h.* Portions of two skulls, very much bleached.
  Mount Victoria, Cameroon Mountains.   Presented by Capt. Burton.

413 *i.* Skull.
  W. Africa.   64, 6, 15, 1.

  2. Tragelaphus decula, *Gray, Cat. Rum. Mam.* p. 50.

16 *a.* Animal, stuffed.
  Abyssinia.   Rüppell.

  3. Tragelaphus sylvatica, *Gray, Cat. Rum. Mam.* p. 50.

649 *a.* Animal, stuffed; male, adult.
  S. Africa.   Kraus.

649 *b.* Animal, stuffed; young male.
  S. Africa.   Verreaux.

649 *c.* Animal, stuffed; female, young.
  S. Africa.   Kraus.

649 *d.* Animal, stuffed; young.
  S. Africa.

649 *e.* Animal, skin; male.
  Africa.

649 *f*. Animal, skin.
Africa.   39, 9, 26, 121.   Verreaux.

649 *g*. Animal, skin; young.
W. Africa.   51, 8, 26, 11.

649 *h*.   Head with the skin on; a young male.
Uganda.   Presented by Captain Speke.

649 *a*. Base of skull, with horns.
Africa.

649 *b*. Horns.
Africa.

649 *c*. Horns.
S. Africa.

649 *d*. Horns.
S. Africa.   48, 7, 11, 3.   Warwick.

649 *i*. Skeleton; male.
S. Africa.   Zool. Society.

649 *l*. Skull; female.
Orange River.   51, 5, 5, 8.

649 *f*. Skull; female.
Orange River.   51, 5, 5, 9.

649 *g*. Skull; male.
Orange River.   51, 5, 5, 10.

649 *h*. Skull; male.
S. Africa.   Gordon Cuming.

649 *j*. Skull; male.
Uganda.   63, 7, 7, 5.   Presented by Captain Speke.

5. PORTAX, *Gray, Cat. Rum. Mam.* p. 51.

1. PORTAX TRAGOCAMELUS, *Gray, Cat. Rum. Mam.* p. 51.

648 *a*. Animal, stuffed; male.
India.   Warwick.

648 *b*. Animal, skin; adult male; in a bad state.
India.   55, 1, 20, 2.   Zool. Society.

648 *c*. Animal, skin; young male; without the horn-sheaths.
India.   47, 5, 17, 20.   Bartlett.

648 *b*. Skeleton.
India.   50, 11, 22, 68.   Zool. Society.

648 *a*. Skull; male.
P. tragocamelus, *Gray, Cat. Mam. B. Mus.* iii. p. 141, fig. 2, 1852.
India.   57, 6, 10, 73.

648 *c*. Skull; male, young.
India.   54, 6, 8, 2.

R

648 *d*. Skull; adult male; lower jaw wanting.
India. 56, 9, 22, 10. Lieut. Abbott.

648 *c*. Skull; female, adult.
India. 60, 4, 22, 1.

Family 14. CAPRIDÆ, *Gray, Cat. Rum. Mam.* p. 51.

1. HEMITRAGUS, *Gray, Cat. Rum. Mam.* p. 51.

1. HEMITRAGUS JEMLAICUS, *Gray, Cat. Rum. Mam.* p. 51; *Cat. Mam. B. M., Ungulata*, tab. xviii. f. 3, 4 (skull and horns).

886 *a*. Animal, stuffed.
Nepal.

886 *b*. Animal, stuffed.
Nepal. 39, 7, 25, 17.   Presented by B. H. Hodgson, Esq.

886 *c*. Animal, skin.
Nepal. 55, 1, 20, 4.   Pres. by H. H. the Maharajah Dhuleep Singh.

886 *p*. Skeleton; male.
India. 62, 3, 19, 16.

886 *b*. Horns, on frontal bone.
Nepal.

886 *d*. Horns; young.
India.

886 *j*. Horns, on frontal bone.
India.

886 *k—m*. Horns, on frontal bone.
Nepal. 45, 1, 8, 180—187—188.   Pres. by B. H. Hodgson, Esq.

886 *a*. Skull, with horns.
Nepal. 42, 4, 10, 12.

886 *c*. Skull, with horns; nasal bones wanting.
India.

886 *f*. Skull.
Nepal. 45, 1, 8, 184.   Presented by B. H. Hodgson, Esq.

886 *i*. Skull, with horns; imperfect.
India.

886 *n*. Skull.
Cashmere.   Presented by Prof. Oldham.

2. KEMAS, *Gray, Cat. Rum. Mam.* p. 51.

1. KEMAS WARRYATO, *Gray, Cat. Rum. Mam.* p. 51; *Cat. Mam. B. M., Ungulata*, tab. xix. f. 1, 2 (skull and horn).

65 *a*. Animal, stuffed; male.
Neilgherries.

65 *b*. Animal, stuffed; female.
Kemas Hylocrius, *Ogilby, P. Z. S.*, 1837, p. 81.
Neilgherries. 55, 12, 24, 291. Zool. Society.

65 *a*. Head, stuffed ; female.
Neilgherries. 42, 2, 24, 2. Presented by R. Partridge, Esq.

65 *b*. Skull ; male.
Neilgherries. Presented by R. Partridge, Esq.

3. ÆGOCEROS, *Gray, Cat. Rum. Mam.* p. 52.

1. ÆGOCEROS PYRENAICA, *Gray, Cat. Rum. Mam.* p. 52.

777 *a*. Animal, stuffed ; male.
Pyrenees. 48, 2, 5, 4.

777 *b*. Animal, stuffed.
Spain. Presented by Prof. Owen.

777 *c*. Animal, skin ; male.
" Capra Schimperi."
Spain. Parzudaki.

777 *a*. Skull ; female.
Pyrenees. 48, 2, 5, 5.

2. ÆGOCEROS CAUCASICA, *Gray, Cat. Rum. Mam.* p. 52.

a. Animal, stuffed ; male.
Aladagh. Warwick.

b. Animal, stuffed ; female.
Aladagh. Warwick.

4. CAPRA, *Gray, Cat. Rum. Mam.* p. 52.

1. CAPRA IBEX, *Gray, Cat. Rum. Mam.* p. 52.

650 *a*. Animal, stuffed ; male.
Alps.

650 *b*. Animal, stuffed ; male.
Alps. Presented by Mrs. A. G. Campbell.

650 *c*. Animal, skin ; young.
650 *g*. Skull of " *c*."
Alps.

650 *d*. Animal, skin ; young.
Syria, Engedi. 64, 8, 17, 15. H. B. Tristram.

650 *e*. Animal, skin.
650 *e*. Skull of " *e*."
Syria, near Engedi. 64, 8, 17, 14. H. B. Tristram.

650 *a*. Horns.
Alps.

650 *b*. Horns.
Alps.

650 *c*. Two left horns.
Alps.

650 *d*. Horns.
Alps.

650 *e*. Skeleton.
Alps.　57, 3, 18, 1.　Warwick.

650 *f*. Skull.
Alps.　57, 3, 18, 2.　Warwick.

650 *h*. Skull, with horns; male.
Alps.

2. Capra sibirica, *Gray, Cat. Rum. Mam.* p. 52.

1359 *a*. Animal, stuffed; male.
Siberia.　St. Petersburgh Museum.

1359 *b*. Animal, stuffed; female.
Siberia.　St. Petersburgh Museum.

1359 *f*. Horns.
India.　56, 5, 6, 118.　　　　　　　Presented by Prof. Oldham.

1359 *g*. One horn, with the tip bent forward.
India.　50, 1, 11, 13.

1359 *c*. Horns.
India.　56, 9, 22, 7.　Abbott.

1359 *e*. Horns.
India.　56, 9, 22, 8.　Abbott.

1359 *a*. Skull, with horns.
"C. altaica."
Altai Mts.　52, 12, 9, 5.　Brandt.

1359 *b*. Skull, with horns.
India.　56, 9, 22, 6.　Abbott.

1359 *d*. Skull, with horns.
Ladank.　　　　　　　　　　Presented by Lieut. Strachey.

3. Capra nubiana, *Gray, Cat. Rum. Mam.* p. 53; *Cat. Mam. B. M.,
Ungulata*, tab. xix. f. 3, 5 (head and horn).

651 *a*. Animal, stuffed; adult male.
N. Africa.

651 *b*. Animal, stuffed; young male.
N. Africa.

651 *c*. Animal, skin; male.
Erzeroom.　Dickson & Ross.　Zool. Society.

651 *a—e*. Horns.
Egypt.　　　　　　　　　　Presented by J. Burton, Esq.

651 *h*. Skull, with horns.
Mount Sinai.

651 *i*. Skull, with horns.
Ibex nubiana, *Gray, Cat. Mam. Brit. Mus.* pt. iii. tab. 19, f. 3, 5,
1852.
N. Africa.　49, 10, 2, 1.

651 *j*. Skull, with horns.
N. Africa.

5. HIRCUS, *Gray, Cat. Rum. Mam.* p. 53.

1. HIRCUS ÆGAGRUS, *Gray, Cat. Rum Mam.* p. 53; *Cat. Mam. B. M.,*
   Ungulata, tab. xx. f. 1, 2, 3 (skull and horns).

653 *a.* Animal, stuffed ; ears pendant.
   Hab. ———— ?

653 *b.* Animal, stuffed.
   Africa.   Presented by Miss Inglis.

653 *c.* Animal, stuffed; with white curly hairs ; ears pendant.
   India. 48, 2, 1, 21.

653 *d.* Animal, skin.
   Nepal. 58, 6, 24, 21.   Presented by B. H. Hodgson, Esq.

653 *e.* Animal, skin.
   Nepal. 58, 6, 24, 30.   Presented by B. H. Hodgson, Esq.

653 *f.* Animal, skin.
   Nepal. 45, 1, 8, 333.

653 *g.* Animal, skin ; large black variety ; ears pendant.
   Hab. ———— ?

653 *a.* Horns ; very large.
   India ?

653 *b.* Horns ; very large.
   India ?

653 *c.* Horns.
   India.? 43, 5, 15, 2.

653 *d.* Horns.
   India ?

653 *e.* Horns.
   India. 56, 9, 22, 8.

653 *f, g.* Skulls.
   Nepal. 45, 1, 8, 171.   Presented by B. H. Hodgson, Esq.

653 *h.* Skull.
   Nepal. 45, 1, 8, 175.   Presented by B. H. Hodgson, Esq.

653 *i.* Skull, *var.* Chappoo.
   "Cupra Chungra."
   Nepal. 45, 1, 8, 177.   Presented by B. H. Hodgson, Esq.

653 *j.* Skull and horns of adult.
   Nepal.   Presented by B. H. Hodgson, Esq.

653 *k.* Skull of " Chappoo."
   Thibet.   Presented by B. H. Hodgson, Esq.

653 *l.* Pair of separate horns.
   Thibet.   Presented by B. H. Hodgson, Esq.

653 *m.* Skull.
   England.

653 *n.* Skull.
   India.

653 *o—r*. Skulls and horns.
Hab. ———— ?

653 *s*. Skull.
Hab. ———— ? Zoological Society.

653 *t*. Skeleton.
Hab. ———— ?

653 *u*. Skull.
Hab. ———— ? 58, 5, 4, 23. Zool. Society.

653 *v*. Skull.
" Scinde goat."
Scinde. 58, 5, 4, 35. Zool. Society.

653 *w*. Skull.
Alps. 59, 9, 6, 112. Günther.

653 *x*. Skull.
Alps. 59, 9, 6. Günther.

653 *y, z*. Skulls, with horns.
" Capra caucasica."
Hab. ———— ?

653 *k²*. Skull.
" Haussa goat."
Bidda. 65, 5, 3, 8. Baikie.

653 *a²—e^n*. Skulls, adult.
Hab. ———— ? 67, 4, 12, 275—277—279—280. Lidth de Jeude.

2. HIRCUS FALCONERI, *Gray, Cat. Rum. Mam.* p. 53.

782 *a*. Animal, skin ; between *H. Falconeri* and common goat.
782 *j*. Skeleton of " *a*."
64, 12, 8, 9. Zool. Gardens.

782 *b*. A furrier's skin.
Pagure. 56, 9, 22, 4. Abbott.

782 *e*. Horns.
Hab. ———— ?

782 *a*. Horns.
" Capra megaceros."
Cashmere. 56, 5, 6, 60. Presented by Prof. Oldham.

782 *b*. Horns.
India ?

782 *d*. Single horn.
India. Presented by H. Falconer, Esq.

782 *f*. Horns.
Pashewar Hills. Abbott.

782 *g*. Horns.
Cashmere.

782 *i*. Single horn.
Himalaya Mountains.

782 *h*. Skull, with horns.
"Capra megaceros," *Hatton, Cal. Journ. Hist.* ii. p. 521, tab. 20.
Cashmere. 56, 5, 6, 60. Presented by Prof. Oldham.

782 *c*. Skull, with horns.
Capra falconeri.
India. Presented by the East India Company.

Family 15. OVIDÆ, *Gray, Cat. Rum. Mam.* p. 54.

1. OVIS, *Gray, Cat. Rum. Mam.* p. 54.

1. OVIS ARIES, *Gray, Cat. Rum. Mam.* p. 54.

670 *a*. Animal, stuffed; four-horned variety, male.

670 *b*. Animal, stuffed; white.
———— 46, 5, 13, 5.

670 *c*. Animal, stuffed. The Guinea sheep.
W. Africa. 50, 11, 30, 12, Whitfield.

670 *d*. Animal, stuffed. Fat-tailed variety.
———— 46, 3, 8, 18.

670 *e*. Animal, stuffed. Female Turkish sheep, with two throat-beards.
Turkey.

670 *f*. Animal, skin. Four-horned sheep.
Nepal. 58, 6, 24, 29. Presented by B. H. Hodgson, Esq.

670 *g*. Animal, skin. Four-horned sheep.
———— Zool. Society.

670 *h*. Animal, skin. Fezzan sheep.
———— 50, 11, 22, 14. Zool. Society.

670 *i*. Animal, skin. Domestic sheep.
Upper Ishadda, Central Africa. 55, 5, 20, 2. Baikie.

670 *j*. Animal, skin. "Barwhal."
Nepal. Presented by B. H. Hodgson, Esq.

670 *a, b*. Horns of many-horned variety.
Hab. ———— ?

670 *c*. Horns, small.
Hab. ———— ?

670 *f, g*. Horns of spiral-horned variety.
Hab. ———— ?

670 q^2. Skeleton. Shanghai or earless ram.
Shanghai. Zool. Society.

670 s^2. Skeleton. Cheviot Scotch ram.
Scotland.

670 w^2. Skeleton. Shetland Island ram.
Shetland Island. 71, 10, 17, 1.

670 t^3. Skeleton. Long-tailed African sheep.
Africa. 67, 7, 8, 2. Zool. Society.

670 *d, e.* Skulls, with horns, of spiral-horned variety.
Hab. ———— ?

670 *h.* Skull, with horns united together at the base.
Hab. ———— ?

670 *i.* Skull.
Nepal.   Presented by B. H. Hodgson, Esq.

670, *j, k.* Skulls.
Nepal.   45, 1, 12, 160.   Presented by B. H. Hodgson, Esq.

670 *l.* Skull.
Nepal.   45, 1, 12, 159.   Presented by B. H. Hodgson, Esq.

670 *m.* Skull; four-horned variety.
Nepal.   Presented by B. H. Hodgson, Esq.

670 *n.* Skull; four-horned variety.
Nepal.   Presented by B. H. Hodgson, Esq.

670 *o.* Skull; four-horned variety.
Nepal.   45, 1, 8, 160.   Presented by B. H. Hodgson, Esq.

670 *p.* Skull; four-horned variety.
Nepal.   Presented by B. H. Hodgson, Esq.

670 *q.* Skull; young. "Cajo."
Nepal.   45, 1, 8, 57.   Presented by B. H. Hodgson, Esq.

670 *r.* Skull, without horn-sheath.
Nepal.   45, 1, 8, 158.   Presented by B. H. Hodgson, Esq.

670 *s.* Skull, with horns united at the base.
Nepal.   45, 1, 8, 161.   Presented by B. H. Hodgson, Esq.

670 *t.* Skull.
Hab. ———— ?   58, 5, 4, 30.   Zool. Society.

670 *u.* Skull, with spiral horns.
Hab. ———— ?   49, 7, 19, 13.

670 *v.* Skull.
England.   49, 3, 3, 1.

670 *w.* Skull, without horns.   Domestic sheep.
England.   47, 12, 11, 7.

670 *x.* Skull.
Hab. ———— ?

670 *y.* Skull.
Hab. ———— ?

670 *z.* Skull.   Cape ram.
Cape of Good Hope.   45, 12, 29, 3.

670 a^2. Skull, without horns.
Hab. ———— ?   58, 5, 4, 29.   Zool. Society.

670 b^2. Skull.
Hab. ———— ?

670 c^2. Skull, without horns.
Hab. ———— ?   58, 5, 4, 28.   Zool. Society.

670 d^2. Skull, without horns.
Hab. ———— ?   58, 5, 4, 27.   Zool. Society.

670 c^2. Skull, without horns. Abyssinian sheep.
Abyssinia. 58, 5, 4, 31. Zool. Society.

670 f^2. Skull, without horns.
Nepal. 58, 6, 24, 130.   Presented by B. H. Hodgson, Esq.

670 g^2. Skull.
India.

670 h^2. Skull. "Barwall sheep."
Nepal. 58, 6, 24, 126,   Presented by B. H. Hodgson, Esq.

670 i^2. Skull. "Barwall sheep."
Nepal. 58, 6, 24, 178.   Presented by B. H. Hodgson, Esq.

670 j^2. Skull.
India.

670 k^2. Skull.
Hab. ———— ?

670 l^2. Skull.
India. 58, 5, 4, 25.

670 m^2. Skull.
Hab.———— ?   58, 5, 4, 24.

670 n^2. Skull, without hairs.
Hab. ———— ? 58, 5, 4, 140.

670 o^2. Skull, without horns.
Hab. ———— ? 58, 5, 4, 134.

670 p. Skull.
Hab. ———— ? Zool. Society.

670 q. Skull.
Alps. 59, 9, 6, 11.

670 r^2 Skull. Iceland ram, four-horned variety.
Iceland.   Presented by Prof. Prosch.

670 u^2. Skull. Hunia sheep.
India.   Presented by the Earl of Derby.

670 v^2. Skull, disarticulated.
England.

670 x^2. Horns. "Barwall."
Nepal. 45, 1, 8, 158.   Presented by B. H. Hodgson, Esq.

670 y^2. Horns. "Barwall."
Nepal. 45, 1, 8, 155.   Presented by B. H. Hodgson, Esq.

670 a^3. Horns. "Barwall."
Hab. ———— ? 45, 1, 8, 153. Lidth de Jeude.

670 b^3. Skull; four-horned variety.
Hab. ———— ? 67, 4, 12, 269. Lidth de Jeude.

670 c^3. Skull; four-horned variety.
Hab. ———— ? 67, 4, 12, 268. Lidth de Jeude.

670 d^3. Skull; four-horned variety.
Hab. ———— ? 67, 4, 12, 267. Lidth de Jeude.

s

670 *e³*. Skull ; four-horned variety.
 Hab. ————— ? 67, 4, 12, 271.   Lidth de Jeude.

670 *f³*. Skull ; spiral-horned variety.
 Hab. ————— ? 67, 4, 12, 266.   Lidth de Jeude.

670 *g³*. Skull ; spiral-horned variety.
 Hab. ————— ? 67, 4, 12, 612.   Lidth de Jeude.

670 *h³*. Skull ; spiral-horned variety.
 Hab. ————— ? 67, 4, 12, 610.   Lidth de Jeude.

670 *i³*. Skull ; spiral-horned variety.
 Hab. ————— ? 67, 4, 12, 265.   Lidth de Jeude.

670 *j³*. Skull ; spiral-horned variety.
 Hab. ————— ? 67, 4, 12, 611.   Lidth de Jeude.

670 *k³*. Skull ; spiral-horned variety.
 Hab. ————— ? 67, 4, 12, 264.   Lidth de Jeude.

670 *l³*. Skull ; spiral-horned variety.
 Hab. ————— ? 67, 4, 11, 263.   Lidth de Jeude.

670 *m³*. Skull.
 Hab. ————— ? 67, 4, 11, 270.   Lidth de Jeude.

670 *n³*. Skull, without horns.
 Hab. ————— ? 67, 4, 11, 272.   Lidth de Jeude.

670 *o³*. Skull, without horns.
 Hab. ————— ? 67, 4, 11, 274.   Lidth de Jeude.

670 *p³*. Skull, without horns.
 Hab. ————— ? 67, 4, 12, 273.   Lidth de Jeude.

670 *q³*. Skull.
 W. Africa. 65, 5, 3, 8.   Baikie.

2. Ovis Polii, *Gray, Cat. Rum. Mam.* p. 54.

667 *a*. Horns.
 Bokhara.   Presented by the E. India Company.

2. CAPROVIS, *Gray, Cat. Rum. Mam.* p. 54.

 1. Caprovis Vignii, *Gray, Cat. Rum. Mam.* p. 55 ; *Cat. Mam. B. M.,
  Ungulata*, tab. xxi. f. 1, 2, 3 (skull and horns).

666 *a*. Animal, stuffed ; young male.
 Ladank. 51, 7, 16, 9. Strachey.

666 *b*. Animal, skin ; female.
 India, Attock hills. 56, 9, 22, 16.   Abbott.

666 *c*. Animal, skin ; young.
 India, Attock hills. 56, 9, 22, 17.   Abbott.

666 *g*. Two separate horns.
 India. 50, 1, 11, 5.

666 *h*. Horns.
 India. 50, 1, 11, 6.

666 *b*. Horns.
N. India. 44, 9, 20. Presented by the E. India Company.

666 *c*. Single horn.
N. India. Presented by the E. India Company.

666 *f*. Horns.
Ladauk. 57, 7, 16, 10. Presented by Capt. Strachey.

666 *a*. Skull, with horns.
N. India. Presented by the E. India Company.

666 *d*. Skull, with horns.
India. 47, 7, 19, 13.

666 *e*. Skull, with horns.
India. 50, 1, 12, 7.

666 *i*. Skull.
India. 56, 9, 22, 12. Abbott.

666 *k*. Skull; young.
Salt Range. 56, 5, 6, 82. Presented by Prof. Oldham.

666 *l*, *m*. Skull; female.
Salt Range. 56, 5, 6, 83—84. Presented by Prof. Oldham.

666 *n*. Skull, without horn-sheaths.
India. 66, 8, 10, 3. Presented by Dr. Falconer.

2. CAPROVIS CYCLOCEROS, *Gray, Cat. Rum. Mam.* p. 55.

1615 *a*. Animal, stuffed; young.
Bred in the Zoological Society's Menagerie. 60, 8, 31, 5.

1615 *b*. Animal, skin.
Hab. ———? 68, 9, 12, 36. Zool. Society.

1615 *a*. Skull, with horns.
India. 56, 5, 6, 81. Presented by Prof. Oldham.

1615 *b*. Skull, with horns.
Ladank. 56, 9, 22, 15. Abbott.

3. CAPROVIS ORIENTALIS, *Gray, Cat. Rum. Mam.* p. 56.

1093 *a*. Animal, stuffed; half-grown male.
Armenia. 47, 9, 22, 2. Presented by the Hon. R. Curzon.

1093 *b*. Animal, skin; young.
Erzeroom. 55, 12, 24, 395. Dickson & Ross. Zool. Society.

1093 *c*. Animal, skin; male.
Erzeroom. 55, 12, 24, 396. Dickson & Ross. Zool. Society.

1093 *b*. Skull.
Hab. ———?

1093 *a*. Skull; male.
Hab. ———? Presented by W. B. Barker, Esq.

1093 *c*. Skull; female.
Erzeroom. 55, 12, 24, 156. Dickson & Ross. Zool. Society.

4. CAPROVIS MUSIMON, *Gray, Cat. Rum. Mam.* p. 56.

1138 *a.* Animal, stuffed ; male.
Hab. ———? 53, 8, 29, 49.   Zool. Society.

1138 *b.* Animal, skin ; female.
Hab. ———? 61, 3, 24, 2.   Zool. Society.

1138 *b.* Skeleton ; male.
Hab. ———? 60, 4, 23, 1.   Zool. Society.

1138 *a.* Skull.
Hab. ———? 55, 12, 26, 162.   Presented by W. Ewer, Esq.

5. CAPROVIS ARGALI, *Gray, Cat. Rum. Mam.* p. 57.

a. Animal, stuffed ; male.
Siberia.

b. Animal, stuffed ; female.
Siberia.

6. CAPROVIS BAMBHERA, *Gray, Cat. Rum. Mam.* p. 57; *Cat. Mam.
B. M., Ungulata,* tab. xxi. f. 4 (skulls and horns).

665 *a.* Animal, skin ; male.
Himalaya Mountains. 59, 8, 7, 1.   Congreve.

665 *b.* Animal, skin ; male.
India.

665 *c.* Animal, skin.
Nepal. 58, 6, 24, 23.   Presented by B. H. Hodgson, Esq.

665 *d.* Animal, skin.
Ladank. Strachey.   Presented by the E. India Company.

665 *a, b.* Horns, adult.
Nepal. 45, 1, 8, 148—150.   Presented by B. H. Hodgson, Esq.

665 *c.* Horns of a young ram.
Nepal. 45, 1, 8, 151.   Presented by B. H. Hodgson, Esq.

665 *l.* Skeleton.
India. 56, 9, 22, 18. Abbott.

665 *d.* Skull of a half-grown ram.
India.

665 *h.* Skull, with horns ; female.
Nepal. 48, 6, 11, 22.   Presented by B. H. Hodgson, Esq.

665 *i.* Skull, with horns ; adult male.
Ladank.   Presented by the E. India Company.

665 *g.* Skull, with horns ; adult male.
Ladank. 51, 12, 22, 2.   Presented by the E. India Company.

665 *k.* Skull ; female.
India. 56, 9, 22, 14. Abbott.

665 *m.* Skull, with horns ; adult male.
Thibet. Steele.

7. CAPROVIS CANADENSIS, *Gray, Cat. Rum. Mam.* p. 57.

1025 *a.* Animal, stuffed ; male.
N. America, Rocky Mountains.   Zool. Society.
Presented by Sir J. Richardson.

1025 *b.* Animal, stuffed ; male.
California.   Presented by the Hudson's Bay Company.

1025 *c.* Animal, stuffed ; young male.
California.   •Presented by the Hudson's Bay Company.

1025 *d.* Animal, skin ; male.
Yellowstone River, Montana.   72, 12, 2, 1.

1025 *e.* Animal, skin.
N. America.   59, 5, 9, 2.   Sir J. Richardson.

1025 *b.* Horns.
Ovis californianus, *Richardson, Zool. Journ.* iv. p. 332.
California.

1025 *a.* Skull, with horns.
N. America.   Zool. Society.

1025 *g.* Skull.
N. America.   67, 2, 23, 4.   Presented by A. G. Dallas, Esq.

1025 *c.* Skeleton ; seven-years-old male.
Vancouver's Is.   59, 12, 29, 7.   Vancouver's Is. Exped. 1848—9.
Presented by Capt. Palisser.

1025 *d, e.* Skeleton ; three- and four-years-old males.
Vancouver's Is.   59, 12, 29, 8—9.   Vancouver's Is. Exped. 1848—9.
Presented by Capt. Palisser.

1025 *f.* Skull, pelvis, and limb-bones.
Vancouver's Is.   59, 12, 29, 10.   Vancouver's Is. Exped.   1848—9.
Presented by Capt. Palisser.

8. PSEUDOIS, *Gray, Cat. Rum. Mam.* p. 57.

1. PSEUDOIS NAHOOR, *Gray, Cat. Rum. Mam.* p. 57; *Cat. Mam. B. M.,*
Ungulata, tab. xxii. f. 3 (head with horns).

668 *a.* Animal, stuffed.
Ovis Bhurrell, *Ogilby, P. Z. S.* 1838, p. 79.
Barinda Pass, N. Chinese Tartary.   Zool. Society.

668 *b.* Animal, stuffed ; male.
India.   57, 7, 29, 3.   Presented by the Linnean Society.

668 *c.* Animal, stuffed ; female.
Nepal.

668 *d.* Animal, skin.
India.   55, 1, 20, 7.   Presented by H. H. Dhuleep Singh.

668 *e.* Animal, skin ; young female.
India.   55, 1, 20, 8.   Presented by H. H. Dhuleep Singh.

668 *f.* Animal, skin ; male.
India.

668 *g*. Animal, skin ; young female.
Nepal.   43, 1, 12, 122.   Presented by B. H. Hodgson, Esq.

668 *g*. Horns.
India.   50, 1, 11, 2.

668 *i*. Horns.
India.   41, 9, 25.

668 *q*. Two horns ; young.
Nepal.   58, 6, 24, 178.   Presented by B. H. Hodgson, Esq.

668 *j*. Horns.
India.   43, 1, 12, 109.

668 *k*. Horns.
India.   52, 6, 25, 5.

668 *l*. Skull, male.
India.   58, 5, 4, 32.   Zool. Society.

668 *m*. Skull ; female.
India.

668 *n*. Skull ; female.
India.   56, 9. 22, 13.   Abbott.

668 *a*. Skull.
Nepal.   48, 6, 11, 23.   Presented by B. H. Hodgson, Esq.

668 *b*. Skull, with horns.
Nepal.   45, 1, 8, 152.   Presented by B. H. Hodgson, Esq.

668 *d*. Skull.
Nepal.   43, 1, 25, 12.   Presented by B. H. Hodgson, Esq.

668 *f*. Skull, without horn-sheaths.
India.   50, 1, 11, 1.

668 *o*. Skull.
India.   Presented by Prof. Oldham.

4. AMMOTRAGUS, *Gray, Cat. Rum. Mam.* p. 58.

   1. AMMOTRAGUS TRAGELAPHUS, *Gray, Cat. Rum. Mam.* p. 58 ; *Cat. Mam. B. M., Ungulatæ*, tab. xxii. f. 1, 2 (skull and horns).

669 *a*. Animal, stuffed ; male.
N. Africa.   Zool. Society.

669 *b*. Animal, stuffed.
Bred in the Zoological Society's Menagerie.   54, 12, 6, 3.

669 *c*. Animal, stuffed.
Bred in the Zoological Society's Menagerie.   48, 8, 21, 139.

669 *a—d*. Horns.
N. Africa ?

669 *i*. Horns.
Tunis.   46, 10, 30, 175.   Fraser.

669 *j*. Horns.
N. Africa ?

669 *k*. Horns, adult.
N. Africa ?

669 *e*. Bones of the body.
N. Africa. 50, 12, 2, 20.

669 *f*. Skeleton ; male.
N. Africa. 56, 12, 30, 1. Zool. Society.

669 *g*. Skeleton ; female.
N. Africa. 58, 1, 12, 3. Zool. Society.

669 *h*. Skeleton ; young.
N. Africa. 60, 12, 30, 17. Zool. Society.

Sub-order 2. DICRANOCERA, *Gray, Cat. Rum. Mam.* p. 59.

Family 16. ANTILOCAPRIDÆ, *Gray, Cat. Rum. Mam.* p. 62.

1. ANTILOCAPRA, *Gray, Cat. Rum. Mam.* p. 63.

1. ANTILOCAPRA AMERICANA, *Gray, Cat. Rum. Mam.* p. 63. Dicranocerus furcifer, *Gray, Cat. Mam. B. M., Ungulata*, tab. xv. f. 1, 2, 3 (skull and horns).

635 *a*. Animal, stuffed ; adult male.
N. America, Rocky Mountains.

635 *b*. Animal, stuffed ; young male.
N. America. 43, 11, 28, 3. Pres. by the Hudson's Bay Company.

635 *c*. Animal, stuffed ; female.
N. America. 72, 5, 6, 3. Zool. Society.

635 *d*. Animal, skin ; young male.
Yellowstone River, Moutana. 73, 12, 12, 2.

635 *e*. Animal, skin ; female.
N. America. 55, 12, 24, 401. Zool. Society.

635 *e*. Single horn.
N. America.

635 *d*. Skeleton ; male ; mounted.
N. America. Zool. Society.

635 *a*. Skull ; female.
Hudson's Bay. Presented by the Hudson's Bay Company.

635 *b*, *c*. Skulls, without the horn-sheaths.
Hudson's Bay.

Sub-order 3. DEVEXA, *Gray, Cat. Rum. Mam.* p. 64.

Family 17. GIRAFFIDÆ, *Gray, Cat. Rum. Mam.* p. 65.

1. GIRAFFA, *Gray, Cat. Rum. Mam.* p. 65.

1. GIRAFFA CAMELOPARDALIS, *Gray, Cat. Rum. Mam.* p. 65. Camelopardalis giraffa, *Gray, Cat. Mam. B. M., Ungulata*, tab. xxiii. f. 1 (skull and horn).

671 *a*. Animal, stuffed ; adult male.
S. Africa. Presented by the Earl of Derby.

671 *b.* Animal, stuffed ; male.
   N. Africa.   Purchased of the Zoological Society.

671 *c.* Animal, stuffed ; young.
   Bred in the Zoological Society's Menagerie.

671 *c.* Skeleton.
   N. Africa.   Zool. Society.

671 *a, b.* Skulls of male and female.
   S. Africa.   Burchell.

Sub-order 4. CAPREOLI, *Gray, Cat. Rum. Mam.* p. 65.

Family 18. ALCADÆ, *Gray, Cat. Rum. Mam.* p. 66.

1. ALCES, *Gray, Cat. Rum. Mam.* p. 66.

   1. ALCES MALCHIS, *Gray, Cat. Rum. Mam.* p. 66; *Cat. Mam. B. M., Ungulata,* t. xxvi. f. 1, 2 (skull and horns), t. xxvii. f. 1 (horn).

703 *a.* Animal, stuffed ; adult female.
   N. Europe.   Presented by the Earl of Derby.

703 *b.* Head, stuffed ; adult male.
   Russia.   Presented by Edward Cayley, Esq.

703 *c.* Skin ; adult female ; without skull.
   Russia.   Zool. Society.

703 *a.* Horns.

———

703 *b.* Two single horns of young.
   ——— 39, 11, 12, 2.

703 *d.* Single horn.
   Sweden, Udoholm. 44, 10, 18, 3.   Pres. by the Earl of Selkirk.

703 *c.* Horns, separate.
   Sweden. Leverian Museum.

703 *j.* Horns.
   St. John's, Newfoundland.

703 *k.* Horns.
   St. John's, Newfoundland.

703 *l.* Horns.
   St. John's, Newfoundland.

703 *o.* Horn, deformed.
   ———   Presented by the Earl of Enniskillen.

703 *p—t.* Horns.

———

703 *f.* Skeleton ; male.
   ——— 50, 11, 22 72.   Zool. Society.

703 *g.* Skeleton ; female.
   ——— 51, 11, 10, 3.   Zool. Society.

703 *h, i.* Skulls.

———

703 *m*. Skull, female.
———— 38, 4, 16, 84.
703 *n*. Skull.
———

Family 19. RANGIFERIDÆ, *Gray, Cat. Rum. Mam.* p. 66.
1. TARANDUS, *Gray, Cat. Rum. Mam.* p. 66.

1. TARANDUS RANGIFER, *Gray, Cat. Rum. Mam.* p. 66; *Cat. Mam. B. M., Ungulata*, t. xxvi. f. 2, 3 (skulls), t. xxvii. f. 2 (horn).

702 *a*. Animal, stuffed; male, half-grown.
N. Europe.   Presented by the Earl of Derby.

702 *b*. Animal, stuffed; adult male.
Hudson's Bay.   Presented by the Hudson's Bay Company.

702 *c*. Animal, stuffed; adult female.
Hudson's Bay.   Presented by the Hudson's Bay Company.

702 *d*. Animal, stuffed; young.
———— 44, 11, 5, 6.

702 *e*. Animal, stuffed; young.
Bred at the Zoological Gardens.   51, 7, 9, 7.

702 *f*. Animal, stuffed; young.
N. Europe.   Bred at Charlton.   Presented by Sir T. Wilson.

702 *q*. Skin of head, with horns.
Newfoundland.   46, 8, 19, 7.

702 *g*. Skin; female, without skull.
———— 43, 11, 28, 4.

702 *e*. Horns, middle-sized.
N. America.   Presented by Capt. Sir J. Franklin.

702 *f*. Horns, small.
N. America.   Presented by Capt. Sir J. Franklin.

702 *g*. Horns, large.
———

702 *h*. Horns, large.
———

702 *i*. Horns.
———

702 *j*. Horns.
N. America.   Presented by Capt. Sir J. Franklin.

702 *k*. Horns, long and slender.
N. America.   Presented by Capt. Sir J. Franklin.

702 *l*. Horns.
N. America.   Presented by Capt. Sir J. Franklin.

702 *m*. Horns, very small.
———

702 *n*. Horns.
———

T

702 *o*. Horns.
    Knowsley.   Presented by the Earl of Derby.

702 *p*. Horns of the year.

---

702 *p—v*. Horns.
    N. W. Coast of America.   Presented by Lieut. Wood & Capt. Kellett.

702 *a*. Skeleton.
    N. Europe. 46, 6, 10, 1.   Presented by Sir S. Wilson.

702 *w*. Skeleton.
    N. America. Dr. J. Rae.

702 *c¹*. Skeleton, male.
    N. Europe. 69, 12, 29, 11. Zool. Society.

702 *x, y, z* & *a—b¹*. A series of skulls with horns.
    N. America.   Presented by A. Murray, Esq.

702 *b*. Skull, large; horns small.

---

702 *c*. Skull and horns, large.

---

702 *d*. Skull and horns, large.

Family 20. CERVIDÆ, *Gray, Cat. Rum. Mam.* p. 67.

1. CERVUS, *Gray, Cat. Rum. Mam.* p. 67.

  1. CERVUS CANADENSIS, *Gray, Cat. Rum. Mam.* p. 68.

690 *a*. Animal, stuffed; adult male.
    N. America.   Presented by Edward Cross, Esq.

690 *b*. Animal, stuffed; adult female.
    N. America.   Presented by W. Tyler, Esq.

690 *c*. Animal, stuffed; young.
    Born at the Surrey Zoological Gardens.

690 *d*. Animal, stuffed; just born.
    Born at the Surrey Zoological Gardens. 53, 8, 29, 43.

690 *a*. Horn, large.
    California, San Diego.   Presented by C. Pentland, Esq.

690 *b*. Horns.
    N. America.

690 *c*. Horns, small.
    N. America.

690 *f*. Horn of the first year.
    ——— 63, 5, 21, 3. Zool. Society.

690 *d*. Horns.
    N. America.

690 *e*. Skeleton.
    ——— Zool. Society.

2. Cervus elaphus, *Gray, Cat. Rum. Mam.* p. 68.

689 *a*. Animal, stuffed; adult male.
Knowsley.   Presented by the Earl of Derby.

689 *b*. Animal, stuffed; adult female.
France. 43, 12, 29.

689 *c*. Animal, stuffed; fawn.
France. 43, 12, 29, 5.

689 *d*. Animal, stuffed; adult male, a pale variety, nearly white.
Alnwick Chase.
            Presented by His Grace the Duke of Northumberland.

689 *a*. Horns.
England.

689 *b*. Horns; left divided into four long cylindrical branches.

689 *c*. Horns, deformed.
Germany.

689 *d*. Horns, deformed.
Germany.

689 *e*. Horns.

689 *f*. Single horn.
Germany.

689 *g, h*. Single horns.
Germany.

689 *i*. Single horn, deformed.
Germany.

689 *j*. Single horn, simple, compressed, notched at the tip.
Germany. 39, 11, 12, 13.

689 *k*. Single horn, deformed.
Germany.

689 *p*. Horns.

689 *q*. Horns, partly fossil.
England.   Presented by Jabez Alliers, Esq.

689 *r*. Single horn.
Asia Minor.   Presented by Lord Arthur Hay.

689 *s*. Single horn, very large.
Crimea.

689 *n*. Skeleton.
———— Zool. Society.

689 *l*. Skull; male.
England.   Presented by the Earl of Derby.

689 *m*. Skull; male.
England. 47, 12, 11, 6.

689 *o*. Skull; female.
    S. Germany. Günther.
689 *t*. Skull; male, with young horns.
    ————— 67, 4, 12, 242. Lidth de Jeude.

3. CERVUS BARBARUS, *Gray, Cat. Rum. Mam.* p. 68.

1042 *a*. Animal, stuffed, just born.
    Born in the Zoological Gardens. 60, 7, 22, 3.
1042 *b*. Animal, stuffed; young.
    N. Africa. 53, 8, 29, 42. Zool. Society.
1042 *b*. Horns.
    Tunis. Zool. Society.
1042 *d, e*. Horns.
    N. Africa. 63, 5, 13, 3—5. Zool. Society.
1042 *a*. Skeleton; female.
    N. Africa. 53, 3, 7, 31. Zool. Society.

4. CERVUS CASHMEERIANUS, *Gray, Cat. Rum. Mam.* p. 68; *Cat. Mam. B. M., Ungulata*, t. xxviii. f. 1 (skull), t. xxvii. f. 3 (horn).

691 *a*. A flat skin, without head or feet.
    Cashmere. 56, 9, 22, 1. Lieut. Abbott.
691 *c*. Single horn; left side.
    India. 57, 1, 22, 7.   Presented by — Money, Esq.
691 *a*. Skull; male.
    Cashmere.   Presented by H. Falconer, M.D.
691 *b*. Skull; female.
    Cashmere.   Presented by H. Falconer, M.D.

5. CERVUS AFFINIS, *Gray, Cat. Rum. Mam.* p. 69.

692 *a*. Skin; adult male.
    Nepal. 57, 12, 14, 2.   Presented by B. H. Hodgson, Esq.
692 *a*. Horns, very large.
    Nepal.   Presented by B. H. Hodgson, Esq.
692 *b*. Horns.
    India.   Presented by Dr. Campbell.
692 *e*. Horns.
    India.   Transferred from the Palæontological Department.
692 *c*. Skull; female.
    ————— 55, 12, 26, 159. Zool. Society.
692 *d*. Skull; male, with rudimentary horns; left side of lower jaw
        wanting.
    India. 66, 8, 10, 5.   Presented by Dr. Falconer.

6. Cervus maral, *Gray, Cat. Rum. Mam.* p. 69.

691 *a.* Animal, stuffed; just born.
Born in the Zoological Gardens. 53, 8, 29, 40.

691 *c—e.* Horns, separate.
Persia. 63, 5, 13, 1—3. Zool. Society.

691 *f.* Horns.
Persia. 65, 7, 8, 3. Zool. Society.

691 *h.* Horns, with the two frontal antlers and the middle snag close
together; right side.
Persia. 58, 5, 4, 11. Zool. Society.

691 *g.* Skeleton; male.
Persia. 66, 8, 6, 12. Zool. Society.

7. Cervus Kopshi, *Swinhoe, MSS.*

1613 *a.* Skin.
1613 *a.* Skull.
Kuikiang. 73, 6, 27, 1. Collected by R. Swinhoe, Esq.

2. PSEUDAXIS, *Gray, Cat. Rum. Mam.* p. 70.

1. Pseudaxis taivanus, *Gray, Cat. Rum. Mam.* p. 70.

1413 *a.* Animal, stuffed; male.
1413 *c.* Bones of the body.
Formosa. Zool. Society.

1413 *b.* Animal, stuffed; fawn.
Formosa. 68, 3, 21, 3. Zool. Society.

1413 *c.* Skin; male, with young horns.
Formosa. 68, 12, 29, 14. Zool. Society.

1413 *d.* Skin; female.
Formosa. 68, 3, 21, 4. Zool. Society.

1413 *a.* Horns.
Island of Formosa. 63, 5, 28, 2. Zool. Society.

1413 *b.* Horns.
Island of Formosa. 65, 1, 30, 1. Zool. Society.

2. Pseudaxis mantchurica, *Gray, Cat. Rum. Mam.* p. 72.

1621 *a.* Animal, stuffed; adult male.
China, Pekin. Swinhoe.

1621 *b.* Skin; female.
1621 *a.* Skull of "*b.*"
Pekin. 61, 6, 2, 2.   Presented by R. Swinhoe, Esq.

1621 *c.* Skin.
Pekin. 61, 6, 2, 3.   Presented by R. Swinhoe, Esq.

1621 *b.* Skull; male, with young horns.
China.

3. Pseudaxis sika, *Gray, Cat. Rum. Mam.* p. 72.

1412 *a*. Animal, stuffed; male; with young horns.
Kanagawa. 64, 2, 30, 3. Zool. Society.

1412 *a*. Horns.
Japan. 63, 5, 281. Zool. Society.

1412 *b*. Horns.
Japan. 60, 12, 121. Zool. Society.

3. DAMA, *Gray, Cat. Rum. Mam.* p. 74.

1. Dama vulgaris, *Gray, Cat. Rum. Mam.* p. 74; *Cat. Mam. B. M.,
Ungulata,* t. xxviii. f. 2, 3 (skulls and horns), t. xxx. f. 1 (horn).

693 *a*. Animal, stuffed; adult male; red, with white spots.
England.

693 *b*. Animal, stuffed; adult male; red, with white spots.
England.

693 *c*. Animal, stuffed; adult male; black, without spots.
England.

693 *d*. Animal, stuffed; adult male; white.
England.

693 *e*. Animal, stuffed; male.
France.

693 *f*. Animal, stuffed; fawn.
France.

693 *g*. Animal, stuffed; fawn.
———— 52, 10, 5, 1.

693 *h*. Furrier's skin, without head.
693 *m*. Skull, with horns; lower jaw wanting.
693 *m*. Skull, with horns. 57, 1, 14, 1 (black variety).
693 *l*. Skull, with horns; lower jaw wanting.
693 *l*. Skull, with horns. 57, 1, 14, 3.
Domesticated, and naturalized about one hundred or one hundred
and fifty years ago, in the Island of Barthda, N. Antigua.
57, 1, 14, 2.   Presented by Charles Darwin, Esq.

693 *i*. Stuffed head, with horns.
Cervus nudipalpebra, *Ogilby,* P. Z. S. 1831.
———— Zool. Society.

693 *c*. Stuffed head, with horns. (Cervus mauricus, *F. Cuv.*)

693 *n, o*. Horns.

693 *a*. Horns.

693 *b*. Horns; a distorted variety.

693 *d*. Horns.

---

693 *e*. Horns; adult.

---

693 *f*. Horns.

---

693 *g*. Horns.

---

693 *x*. Forty-six horns, more or less deformed.
New Forest. 50, 2, 5.   Presented by Mrs. Smyth.

693 *i*. Skeleton.
England. 47, 12, 11, 1.

693 *h*. Skull; adult male.
England.

693 *k*. Skull; young female.
England. 54, 6, 8, 3.

693 *q*. Skull; female.
S. Germany. Günther.

693 *r*. Skull; female.
Europe. 58, 5, 4, 22. Zool. Society.

693 *s*. Skull; male.
Europe. 67, 4, 12, 234. Lidth de Jeude.

693 *t*. Skull, with young horns.
———— 67, 4, 12, 235. Lidth de Jeude.

693 *u*. Skull, with young horns.
———— 67, 4, 12, 236. Lidth de Jeude.

693 *v*. Skull, with young horns.
———— 67, 4, 12, 237. Lidth de Jeude.

693 *w*. Skull; female.
———— 67, 4, 12, 241. Lidth de Jeude.

4. PANOLIA, *Gray, Cat. Rum. Mam.* p. 75.

1. PANOLIA ELDII, *Gray, Cat. Rum. Mam.* p. 75; *Cat. Mam. B. M.,
Ungulata*, t. xxix. f. 1, 2 (skull), t. xxx. f. 2 (horn).

695 *a*. Animal, stuffed; adult male, with diseased horns.
Cervus dimorphe, *Hodgson.*
India.   Presented by B. H. Hodgson, Esq.

695 *b*. Skin; young female.
695 *o*. Bones of the body of "*b*."
India. 68, 12, 28, 9. Zool. Society.

695 *c*. A flat skin, without head or feet; female.
China, Hainan. 70, 2, 10, 32. Swinhoe.

695 *d*. A flat skin, without feet.
China, Hainan. 70, 2, 10, 28. Swinhoe.

695 *e.* Skin; fawn.
   China, Hainan.   70, 2, 10, 29.   Swinhoe.

695 *f.* A flat skin, without the head or feet; very young.
   China, Hainan.   70, 2, 10, 30.   Swinhoe.

695 *a.* Horns.
   India.

695 *b.* Horns.
   India.   Presented by B. H. Hodgson, Esq.

695 *c.* Horns.
   India.   Presented by B. H. Hodgson, Esq.

695 *g.* Horns.
   India.

695 *i.* Horns, very large.
   India.

695 *l.* Single horn.
   Nepal.   Presented by B. H. Hodgson, Esq.

695 *m.* Horns.
   Siam and Camboja.   Sir R. Schomburgh.

695 *n.* Horns; young.
   Siam and Camboja.   Presented by Sir R. Schomburgh.

695 *w.* Horns.
   India ?   Transferred from the Palæontological Department.

695 *p—v.* A series of young horns.
   Formosa.   70, 2, 10.   Swinhoe.

695 *d.* Skull, with horns.
   India.

695 *e.* Skull, with horns.
   India.

695 *f.* Skull, with horns.
   India.

   2. Panolia platycercus, *Gray, Cat. Rum. Mam.* p. 75.

695 *h.* Single horn.
   India.

5. RUCERVUS, *Gray, Cat. Rum. Mam.* p. 75.

   1. Rucervus Duvaucellii, *Gray, Cat. Rum. Mam.* p. 76; *Cat.
      Mam. B. M., Ungulata,* t. xxix. f. 4 (skull), t. xxx. f. 3 (horn).

694 *a.* Animal, stuffed; adult male.
694 *d.* Skull of "*a.*"
   India.   Presented by the Earl of Derby.

694 *b.* Animal, stuffed; fawn.
   India.   63, 12, 3, 7.   Zool. Society.

694 *c.* Animal, stuffed; fawn.
   Born in the Zoological Gardens.   64, 8, 17, 3.

694 *d*. Skin; female.
India. 55, 12, 24, 400. Presented by the Earl of Derby.

694 *f*. Horns.
Nepal. 45, 1, 8, 131. Presented by B. H. Hodgson, Esq.

694 *g*. Horns.
Nepal. 45, 1, 8, 130. Presented by B. H. Hodgson, Esq.

694 *i*. Horns.
India. 63, 5, 28, 3. Zool. Society.

694 *j*. Single horn, left side; young.
Nepal. 58, 6, 24, 176. Presented by B. H. Hodgson, Esq.

694 *k*. Horns; young.

---

694 *h*. Skeleton; skull, with only the left horn developed.
——— 60, 4, 23, 3. Zool. Society.

694 *a*. Skull, with horns.
Nepal. 45, 1, 8, 128. Presented by B. H. Hodgson, Esq.

694 *b*. Skull, with horns.
Nepal. Presented by B. H. Hodgson, Esq.

694 *c*. Skull; female.
Nepal. 45, 1, 8, 197. Presented by B. H. Hodgson, Esq.

694 *e*. Skull, with horns.
Nepal. 45, 1, 8, 129. Presented by B. H. Hodgson, Esq.

694 *l*. Skull; female.
Nepal. 45, 1, 8, 200. Presented by B. H. Hodgson, Esq.

---

2. RUCERVUS CAMBOJENSIS, *Gray, Cat. Rum. Mam.* p. 76.

1463 *a*. Horns, on frontal bone.
Cervus Schomburghii, *Blyth, P. Z. S.* 1863, p. 155, fig.
Siam and Camboja. Sir R. Schomburgh.

1463 *b*. Horns, separate.
Siam and Camboja. Sir R. Schomburgh.

1463 *c*. Horns.

1463 *d—f*. Horns, separate.
C. Schomburghii, *Blyth, P. Z. S.* 1867, p. 836, fig.
67, 8, 20, 1—3. Blyth.

1463 *g*. Horns, on frontal bone.
Cervulus cambojensis, *Gray, P. Z. S.* 1861, p. 138.
Camboja. 61, 4, 12, 18. Mouhot.

6. RUSA, *Gray, Cat. Rum. Mam.* p. 76.

1. RUSA ARISTOTELIS, *Gray, Cat. Rum. Mam.* p. 76; *Cat. Mam. B. M., Ungulata*, t. xxxi. f. 12 (skull), t. xxx. f. 4 (horn).

699 *a*. Animal, stuffed; adult male.
India. Presented by B. H. Hodgson, Esq.

U

699 *b*. Animal, stuffed; male.
699 *q²*. Skull of "*b*."
  Ceylon. 58, 4, 22, 1.   Zool. Society.

699 *c*. Skin; male; with young horns.
  Nepal. 58, 6, 24, 19.   Presented by B. H. Hodgson, Esq.

699 *d*. Skin; male; with young horns.
  Nepal. 58, 6, 24, 20.   Presented by B. H. Hodgson, Esq.

699 *e*. A flat skin, without head or feet.
  India.   Presented by Gen. Hardwicke.

699 *j*. Head and horns, with the skin.
  India.   Presented by Edw. Cross, Esq.

699 *k*. Head of female.
  Himalaya.   43, 1, 12, 107.   Presented by B. H. Hodgson, Esq.

699 *a*. Single horn.
  India.   Presented by Edw. Cross, Esq.

699 *b*. Horns, on frontal bone.
  India.

699 *c*. Two single horns.
  India.

699 *d*. Horns, on frontal bone.
  India.

699 *e*. Horns, on frontal bone.
  India.

699 *f*. Horns, on frontal bone.
  India.   Presented by Mrs. Wright.

699 *g*. Horns, on frontal bone.
  India.

699 *h*. Horns, on frontal bone.
  India.

699 *i*. Single horn.
  India.

699 *o*. Horns.
  Nepal.   Presented by B. H. Hodgson, Esq.

699 *p*. Horns.
  Nepal.   Presented by B. H. Hodgson, Esq.

699 *q*. Horns.
  Nepal.   Presented by B. H. Hodgson, Esq.

699 *r*. Horns.
  Nepal.   Presented by B. H. Hodgson, Esq.

699 *s*. Single horn of the first year.
  India.

699 *t*. Horns.
  India.

699 *u*. Horns.
  India.

699 *v*. Horns.
Himalaya.   Presented by the Rev. R. Everest.

699 *w*. Horns, apex simple.
India.

699 *x*. Horns, separate.
India.

699 *y*. Horns, very large.
Nepal.   45, 1, 8, 107.   Presented by B. H. Hodgson, Esq.

699 *z*. Horns.
Nepal.   Presented by B. H. Hodgson, Esq.

699 *a—c²*. Three pairs of large horns.
Nepal.   Presented by B. H. Hodgson, Esq.

699 *d²*. Horns.
Nepal.   Presented by B. H. Hodgson, Esq.

699 *e²*. Horns.
Nepal.   Presented by B. H. Hodgson, Esq.

699 *t²*. Horns.
Himalaya Mountains.

699 *u²*. Single horn, very large.
Himalaya Mountains.

699 *v²*. Horns.
Nepal.   Presented by B. H. Hodgson, Esq.

699 *w²—x²*. Horns.
India.   58, 5, 4, 13.   Zool. Society.

699 *z²*. Horns, on frontal bone.
Assam.   67, 5, 15, 3.

699 *a³*, *b³*. Horns, on frontal bone.
Assam.   67, 5, 15, 4—5.

699 *c³*. Single horn, left side, very large and heavy.
Ceylon.   Capt. Lewis.

699 *f²*. Horns.
Nepal.   Presented by B. H. Hodgson, Esq.

699 *g²*. Horns.
Nepal.   Presented by B. H. Hodgson, Esq.

699 *h²*. Horns.
India.   Presented by B. H. Hodgson, Esq.

699 *i²*. Horns.
India.   Presented by B. H. Hodgson, Esq.

699 *j²*. Horns, the beam not branched,
India.

699 *k²*. Horns.
India.   Presented by B. H. Hodgson, Esq.

699 *s²*. Horns.
Himalaya Mountains.

699 y^2. Skeleton; male.
  India.

699 l^2. Skeleton, imperfect.
  Nepal.   Presented by B. H. Hodgson, Esq.

699 m^2. Skeleton.
  India.

699 n^2. Skull; adult female.
  India.

699 o^2, p^2. Skulls; young females.
  India.

699 c^3. Skull, with horns.
  Assam. 67, 5, 15, 6.

699 d^3. Skull; young.
  India. 66, 8, 10, 6.   Presented by Dr. Falconer.

699 p^2. Skull, with horns.
  India.   Presented by B. H. Hodgson, Esq.

699 l. Skull of young male.
  Nepal. 45, 1, 8, 198.   Presented by B. H. Hodgson, Esq.

699 m. Skull; adult female.
  Nepal. 45, 1, 8, 199.   Presented by B. H. Hodgson, Esq.

2. RUSA HIPPELAPHUS, *Gray, Cat. Rum. Mam.* p. 77.

697 a. Animal, stuffed; female.
  India. Zool. Society.

697 b. Animal, stuffed; young.
  India.

697 c. Animal, stuffed; male; with young horns.
  Cervus Tunjuc, *Vigors and Horsfield, Raffles' Memoir*, p. 645.
  Sumatra. 51, 9, 18, 10.   Presented by Sir Stamford Raffles.

697 d. Animal, stuffed; adult female.
  Cervus Tunjuc, *Vigors and Horsfield, Memoirs*, p. 645.
  Sumatra. 51, 9, 18, 11.   Presented by Sir Stamford Raffles.

697 e. A flat skin, without head or feet; female.
  China, Hainan. 70, 2, 10, 31. Swinhoe.

697 f. A flat skin, without head or feet; male.
  China, Hainan. 70, 2, 10, 30. Swinhoe.

697 g. Horns.
  Nepal.   Presented by B. H. Hodgson, Esq.

697 i, j. Horns, on frontal bone.
  India.

697 k. Horns, on frontal bone.
  India.

697 m. Horns, on frontal bone.
  India. 45, 11, 8, 34.

697 *o*. Horns, on frontal bone.
India.

697 *p*. Horns.
Nepal. 45, 1, 8, 114. Presented by B. H. Hodgson, Esq.

697 *q*. Horns.
India. 45, 12, 27, 3.

697 *u*. Horns.
Axis Pennantii, *Gray, List Mam. B. M.* 1843, p. 180.
India.

697 *v*. Horns.
India.

697 *h*. Skull, with horns.
India.

3. RUSA MOLUCCENSIS, *Gray, Cat. Rum. Mam.* p. 77.

1427 *a*. Animal, stuffed; female.
Java.

1427 *b*. Animal, stuffed; female.
Java? 51, 12, 2, 1. Zool. Society.

1427 *c*. Skin; female.
Java? 63, 12, 28, 13. Zool. Society.

1427 *d*. Skin of the head.
1427 *d*. Skull of "*d*."
Batchian. 61, 12, 11, 28. Wallace.

1427 *a*. Skeleton; young male.
Java? 64, 12, 20, 2.

1427 *b*. Skeleton.
Java?

1427 *c*. Skeleton.
Java? 65, 12, 8, 30.

4. RUSA EQUINUS, *Gray, Cat. Rum. Mam.* p. 77.

Animal, stuffed; young male.
Assam?

Skin; adult male.
Assam?? 67, 7, 8, 24. Zool. Society.

Skin; male.
68, 3, 21, 5.

5. RUSA SWINHOII, *Gray, Cat. Rum. Mam.* p. 77.

1414 *a*. Animal, stuffed; young male.
Formosa.

1414 *b*. Skin; male.
Formosa. 68, 3, 21, 24.

1414 *a*. Horns of the first year.
　Zool. Society Gardens.　63, 5, 13, 8.

1414 *d*. Horns; adult.
　Cervus Swinhoii, *Swinhoe, P. Z. S.* 1870, p. 647, fig. 5.
　Formosa.　70, 2, 10, 77.　Swinhoe.

1414 *e*. Horns.
　Formosa.　70, 2, 10, 78.　Swinhoe.

1414 *b*. Skull; male, without horns.
　Formosa.　70, 2, 10, 69.　Swinhoe.

1414 *c*. Skull; young female.
　Formosa.　70, 2, 10, 70.　Swinhoe.

　　6. Rusa marianus, *Gray, Cat. Rum. Mam.* p. 78.

655 *a*. Horns, on frontal bone.
　Philippine Islands.　47, 3, 4, 22.
　　　　　　　　　　Presented by Capt. Sir Edward Belcher, K.C.B.

655 *e*. Horns, separate.
　———　46, 7, 5, 2.

655 ? *f*. Horns, front snag very long.
　———　64, 10, 1, 4.

655 *b*. Skull, with horns.

655 *c*. Skull, with horns.

655 *d*. Skull, with horns.
　———　53, 10, 6, 2.

　　7. Rusa Peronii, *Gray, Cat. Rum. Mam.* p. 78.

a. Animal, stuffed; female.
　Timor.　67, 1, 30, 2.　Zool. Society.

　　8. Rusa Kuhlii, *Gray, Cat. Rum. Mam.* p. 79.

a. Animal, stuffed; male, with adult horns.
　Java.　Leyden Museum.

b. Animal, stuffed; male, with young horns.
　Java.　46, 2, 16, 1.

c. Skin; female.
　Java.　71, 3, 3, 4.　Zool. Society.

　　7. HYELAPHUS, *Gray, Cat. Rum. Mam.* p. 79.

　　1. Hyelaphus porcinus, *Gray, Cat. Rum. Mam.* p. 79; *Cat. Mam.
　　　B. M., Ungulata,* t. xxxii. f. 1 (skull), t. xxxiv. f. 2 (horn).

698 *a*. Animal, stuffed; male.
　India.　47, 5, 17, 22.

698 *b.* Animal, stuffed; male.
India.

698 *c.* Animal, stuffed; very young.
Born in the Zoological Gardens. 58, 5, 26, 5.

698 *d.* Animal, stuffed; just born.
Born in the Zoological Gardens. 65, 5, 19, 19.

698 *e.* Animal, stuffed; male, with young horns.
India. 48, 8, 18, 8.

698 *f.* Skin; female.
Ceylon. 58, 12, 16, 2. Zool. Society.

698 *g.* Skin; female.
Nepal. 58, 6, 24, 113.   Presented by B. H. Hodgson, Esq.

698 *c.* Horns, separate.
Nepal.   Presented by B. H. Hodgson, Esq.

698 *f.* Horns, on frontal bone.
India. 45, 1, 8, 120.   Presented by B. H. Hodgson, Esq.

698 *g—i.* Horns.
India.   Presented by Gen. Hardwicke.

698 *j.* Horns, not forked at the tip.
India.

698 *l—o.* Horns.
India. 67, 12, 24, 4—7.

698 *t.* Horns, separate.
Nepal. 45, 1, 8, 134.   Presented by B. H. Hodgson, Esq.

698 *u.* Horns, separate.
Nepal. 45, 1, 8, 133.   Presented by B. H. Hodgson, Esq.

698? *s.* Skeleton; old female.
Ceylon. 58, 12, 16, 2. Zool. Society.

698 *v.* Skull, with horns.
Assam. 67, 5, 20, 7.

698 *a.* Skull; female.
Nepal. 45, 1, 8, 194.   Presented by B. H. Hodgson, Esq.

698 *b.* Skull, with horns.
Nepal. 45, 1, 8, 123.   Presented by B. H. Hodgson, Esq.

698 *d.* Skull, with horns.
India.

698 *e.* Skull, with horns.
India.

698 *k.* Skull, with horns.
India.

698 *p.* Skull; male.
India. 52, 2, 28, 6.

698 *q.* Skull; male.
India. 56, 5, 6, 62.   Presented by Professor Oldham.

698 *r*. Skull; male, without horns.
India.   58, 5, 4, 19.   Zool. Society.

8. AXIS, *Gray, Cat. Rum. Mam.* p. 79.

   1. AXIS MACULATA, *Gray, Cat. Rum. Mam.* p. 80; *Cat. Mam. B. M.,*
     Ungulata, t. xxxi. f. 3, 4 (skull), t. xxxiv. f. 1 (horn).

697 *h*. Animal, stuffed; male.
India.

697 *a*. Animal, stuffed; female, adult.
India.

697 *b*. Animal, stuffed; female, half grown.
India.   51, 3, 14, 3.   Zool. Society.

697 *c*. Animal, stuffed; young.
51, 2, 17, 4.   Zool. Society.

697 *d*. Animal, stuffed; young.
India.   46, 4, 28, 4.

697 *e*. Animal, stuffed; just born.
Born in Zoological Gardens.   59, 2, 3, 7.

697 *f*. Skin, in a bad state.
Nepal.   58, 6, 24, 16.   Presented by B. H. Hodgson, Esq.

697 *g*. Skin; two years old.
In the Tarrai, near Bamour.   48, 8, 14, 17.   Capt. Boys.

697 *h*. Skin; young.
Nepal.   58, 6, 24, 18.   Presented by B. H. Hodgson, Esq.

697 *a*. Horns, on frontal bone.
India.

697 *b*. Horns, on frontal bone.
India.

697 *c*. Horns, on frontal bone.
India.

697 *e*. Horns, on frontal bone.
Nepal.   45, 1, 8, 120.   Presented by B. H. Hodgson, Esq.

697 *f*. Horns, on frontal bone, deformed.
Nepal.   45, 1, 8, 121.   Presented by B. H. Hodgson, Esq.

697 *n*. Horns, on frontal bone.
Nepal.   45, 1, 8, 332.   Presented by B. H. Hodgson, Esq.

697 *l*. Horns, on frontal bone.
India.   50, 1, 11, 20.

697 *q*. Skeleton; male.
India.   51, 11, 10, 7.   Zool. Society.

697 *s*. Skeleton; male.
India.   57, 2, 24, 11.   Zool. Society.

697 *d*. Skull and horns.
Nepal.   69, 1, 8, 119.   Presented by B. H. Hodgson, Esq.

697 *r*. Skull; female.
India. 55, 12, 26, 158. Zool. Society.

697 *t*. Skull; male.
S. Germany (introduced). 59, 9, 6, 104. Günther.

697 *w*. Skull; male, with young horns.
India. 67, 4, 12, 241. Lidth de Jeude.

9. CAPREOLUS, *Gray, Cat. Rum. Mam.* p. 80.

1. CAPREOLUS CAPRÆA, *Gray, Cat. Rum. Mam.* p. 81; *Cat. Mam.
B. M., Ungulata*, t. xxxiii. f. 1 (skull), t. xxxiv. f. 3 (horn).

688 *a*. Animal, stuffed; adult male in winter.
France.

688 *b*. Animal, stuffed; young male.
Scotland.   Presented by the Earl of Derby.

688 *c*. Animal, stuffed; adult female in summer.
France.

688 *d*. Animal, stuffed; adult female in winter.
Scotland.

688 *e*. Animal, stuffed; fawn.
Europe.

688 *f*. Skin; male, in summer.
Europe. 58, 6, 18, 8. Zool. Society.

688 *a*. Horns, on frontal bone; adult.
Scotland.

688 *b*. Horns; second year.
Scotland.

688 *c*. Single horn.
Scotland.

688.*d*. Horns.
Scotland.   Presented by Gen. Hardwicke.

688 *e*. Horns, very much diseased.
Europe.

688 *g*. Horns.
Europe. 47, 7, 19, 12.

688 *h*. Single horn.
Europe.

688 *i*. Two single horns; young.
Europe.

688 *j*. Skeleton.
Europe. 55, 12, 26, 247. Zool. Society.

688 *f*. Skull; female.
Europe. 47, 7, 19, 12.

688 *k*. Skull; male.
S. Germany. 59, 9, 6, 107. Günther.

x

688 *l*. Skull; female.
  S. Germany. 59, 9, 6, 108. Günther.

688 *m*. Skull; female.
  S. Germany. 59, 9, 6, 109. Günther.

688 *n*. Skull; very young.
  S. Germany. 59, 9, 6, 110. Günther.

688 *o*. Skull; male.
  Europe.

688 *p*. Skull; half-grown.
  Europe.

688 *q*. Skull; adult male.
  Europe. 67, 4, 12, 225. Lidth de Jeude.

688 *r*. Skull; male.
  Europe. 67, 4, 12, 626. Lidth de Jeude.

688 *s*. Skull; male.
  Europe. 67, 4, 12, 227. Lidth de Jeude.

688 *t*. Skull, without the lower jaw; the right horn only developed.
  Europe. 67, 4, 12, 238. Lidth de Jeude.

688 *u*. Skull; young male.
  Europe. 67, 4, 12, 229. Lidth de Jeude.

688 *v*. Skull, without the lower jaw; young male.
  Europe. 67, 4, 12, 630. Lidth de Jeude.

698 *w*. Skull; young male.
  Europe. 67, 4, 12, 631. Lidth de Jeude.

688 *x*. Skull; adult female.
  Europe. 67, 4, 12, 232. Lidth de Jeude.

688 *y*. Skull; female.
  Europe. 67, 4, 12, 233. Lidth de Jeude.

2. CAPREOLUS PYGARGUS, *Gray, Cat. Rum. Mam.* p. 82.

a. Animal, stuffed; adult male.
  China. 70, 7, 18, 14. Swinhoe.

b. Animal, stuffed; adult male.
  Siberia. 42, 3, 13, 1.

c. Animal, stuffed; adult female.
  Siberia. 42, 3, 13, 2.

10. ELAPHURUS, *Gray, Cat. Rum. Mam.* p. 82.

1. ELAPHURUS DAVIDIANUS, *Gray, Cat. Rum. Mam.* p. 82.

1538 *a*. Skin; male.
  N. China. 72, 12, 31, 3. Zool. Society.

1538 *b*. Skin; male.
1538 *a*. Skeleton; male.
  China. 70, 6, 22, 14. Zool. Society.

11. CARIACUS, *Gray, Cat. Rum. Mam.* p. 82; *Cat. Mam. B. M.,* *Ungulata,* t. xxxiv. f. 4 (horn).

   1. CARIACUS VIRGINIANUS, *Gray, Cat. Rum. Mam.* p. 83; *Cat. Mam. B. M., Ungulata,* t. xxxiii. f. 2, 3 (skull and horns).

681 *a.* Animal, stuffed; male.
  N. America.

681 *b.* Animal, stuffed; female.
  N. America.   Presented by the Earl of Derby.

681 *c.* Animal, stuffed; young male.
  N. America.   Presented by the Earl of Derby.

681 *d.* Animal, stuffed; young female.
  N. America. 51, 7, 9, 4.

681 *e.* Animal, stuffed; young male.
  N. America. 51, 7, 9, 3. Zool. Society.

681 *a.* Horns.
  N. America.

681 *b.* Horns.
  N. America. 39, 11, 12, 5.

681 *c.* Horns.
  N. America.

681 *m.* Horns.
  N. America.

681 *n.* Horns.
  N. America.

681 *p.* Horns.
  N. America.

681 *h.* Skeleton, mounted.
  Knowsley.   Presented by the Earl of Derby.

681 *q.* Skeleton; young male.
  N. America. 50, 11, 22, 55. Zool. Society.

681 *r.* Skeleton; male.
  N. America. 51, 11, 10, 6. Zool. Society.

681 *s.* Skull; young.
  N. America.   Presented by the Earl of Derby.

   2. CARIACUS LEUCURUS, *Gray, Cat. Rum. Mam.* p. 83.

683 *a.* Animal, stuffed; adult male.
  N. America. Zool. Society.

683 *b.* Animal, stuffed; adult male, with young horns.
  N. America.

681 *d.* Horns, on frontal bone.
  N. America.

681 *e.* Horns, on frontal bone.
  N. America.

681 *e.* Horns, on frontal bone.
N. America.

681 *f.* Single horn, with many branches.
N. America.

681 *g.* Single horn, with many branches.
N. America.

683 *a.* Skull; adult male.
Columbia River. 45, 7, 4, 5.   Pres. by the Hudson's Bay Company.

683 *b.* Skull; young female.
Columbia River. 45, 7, 4, 5.   Pres. by the Hudson's Bay Company.

3. CARIACUS SIMILIS, *Gray, Cat. Rum. Mam.* p. 83.

681 *a.* Animal, stuffed; female.
Hab. ————? 53, 8, 29, 46.   Zool. Society.

681 *w.* Skull; male, without lower jaw.
Hab. ————? 59, 9, 28, 6.

681 *x.* Skull; male.
Hab. ————? 59, 9, 28, 5.

4. CARIACUS MEXICANUS, *Gray, Cat. Rum. Mam.* p. 84.

1374 *a.* Skin; young male.
Costa Rica. 65, 5, 18, 37.   Salvin.

1374 *b.* Skin; male.
Honduras. 58, 6, 18, 5.

1374 *c.* Skin; adult female.
Costa Rica. 65, 5, 18, 36.   Salvin.

1374 *a.* Horns; adult. (Pl. xxxvi. f. 2.)
Vera Cruz. 56, 12, 4, 1.   Sallé.

1374 *b.* Horns; adult.
Vera Cruz. 56, 12, 4, 2.   Sallé.

1374 *d—g.* Separate horns.
S. Mexico, Oaxaca, "11,000 feet." 58, 6, 2, 18.

1374 *h, i.* Horns; young.
S. Mexico, Oaxaca. "11,000 feet." 58, 6, 2, 18.

1374 *j.* Horns.
Mexico? 47, 7, 19, 14.

1374 *j.* Horns, on frontal bone.
Costa Rica. 67, 8, 23, 2.

1374 *k.* Single horn, on frontal bone.
S. America. 50, 1, 11, 22.

1374 *c.* Skull, with horns. (Pl. xxxvi.)
S. Mexico, Oaxaca, "11,000 feet." 58, 6, 2, 18.

5. CARIACUS LEPTOCEPHALUS, *Gray, Cat. Rum. Mam.* p. 85.

a. Skull; male. (Pl. xxxvii.)
S. America. 55, 12, 26, 160. Zool. Society.

12. EUCERVUS, *Gray, Cat. Rum. Mam.* p. 85.

1. EUCERVUS MACROTIS, *Gray, Cat. Rum. Mam.* p. 86.

1619 *a.* Skin; adult male.
Yellowstone River, Montana. 72, 12, 12, 3.

1619 *b.* Skin; adult female.
Yellowstone River, Montana. 72, 12, 12, 4.

1619 *c.* Skin; female.
Fort Colville. 63, 2, 24, 41.　　　　　Presented by J. K. Lord, Esq.

1619 *a.* Skull; male. (Pl. xxxviii.)
N. America.

2. EUCERVUS COLUMBIANUS, *Gray, Cat. Rum. Mam.* p. 86.

681 *i.* Skull and horns; male. (Pl. xxxix.)
N. America. 45, 7, 4, 3.　　　　Pres. by the Hudson's Bay Company.

681 *j.* Skull; female.
N. America. 45, 7, 4, 3.　　　　Pres. by the Hudson's Bay Company.

681 *k.* Horns.
N. America. 71, 4, 8, 1.

3. EUCERVUS PUSILLA.

a. Skull, without intermaxillary bones or lower jaw, and with subulate
unequal horns, with narrow oblong very deep tear-pit.

N. America. 68, 2, 13, 2. Dr. Robert Brown.

13. BLASTOCERUS, *Gray, Cat. Rum. Mam.* p. 87.

1. BLASTOCERUS PALUDOSUS, *Gray, Cat. Rum. Mam.* p. 87.

687 *a.* Horns.
S. America.

687 *b.* Horns.
S. America.

687 *c.* Horns.
S. America. 71, 6, 20, 2.

687 *d.* Horns.
S. America. 70, 11, 4, 1.

2. BLASTOCERUS CAMPESTRIS, *Gray, Cat. Rum. Mam.* p. 87; *Cat.
Mam. B. M., Ungulata,* t. xxxv. f. 2, 3 (skull and horns).

686 *a.* Animal, stuffed; adult male.
S. America.　　　　　　　Presented by Charles Darwin, Esq.

686 *b*. Animal, stuffed ; young male.
   N. Patagonia.   Pres. by Sir W. Burnett & Capt. Fitzroy, R.N.

686 *c*. Animal, stuffed ; female.
   N. Patagonia.   Pres. by Sir W. Burnett & Capt. Fitzroy, R.N.

686 *d*. Animal, stuffed ; fawn.
   Bolivia.   46, 7, 28, 40.

686 *e*. Animal, stuffed ; fawn.
   N. Patagonia.   Pres. by Sir W. Burnett & Capt. Fitzroy, R.N.

686 *f*. Animal, stuffed ; fawn.
   Born in the Zoological Gardens.   60, 11, 4, 20.

686 *g*. Skin ; male.
   S. America.

686 *a*. Horns, on frontal bone.
   N. Patagonia.   Capt. Fitzroy's Expedition.   From the Haslar
      Hospital.

686 *b*. Imperfect skull, with horns.
   Columbia.   52, 2, 26, 11.

686 *c*. Skeleton, imperfect.
   La Plata.   54, 8, 16, 1.

686 *d*. Skull, with horns.
   La Plata.   54, 8, 16, 2.

686 *e*. Skull, with horns.
   La Plata.   54, 8, 16, 3.

686 *f*. Horns, on frontal bone.
   La Plata.   54, 8, 16, 4.

686 *g*. Imperfect skull, with horns.
   La Plata.   54, 8, 16, 5.

686 *i*. Imperfect skull, with horns.
   La Plata.   54, 8, 16, 7.

686 *j*. Skeleton ; female.
   S. America.   61, 11, 15, 2.   Zool. Society.

686 *k*. Skull ; male.
   S. America.   58, 6, 9, 17.   Zool. Society.

3. BLASTOCERUS SYLVESTRIS, *Gray, Ann. & Mag. N. Hist.*, 1873.

a. Brain-case of the skull ; with horns short, slender, smooth (7 in.
      high), forked above. Front snag elongate, projecting forwards
      and suddenly curved upwards, with a slight tubercle on the
      middle of the lower edge, and a small snag rather below it on
      the inner side of the upper edge.   (Pl. xl.)
Blastocerus sylvestris, *Gray*, l. c. 1873.
   Brazils. "Wood-deer."   72, 2, 11, 1.
                           Presented by the Rev. G. T. Hudson.

14. FURCIFER, *Gray, Cat. Rum. Mam.* p. 88.

1. FURCIFER ANTISIENSIS, *Gray, Cat. Rum. Mam.* p. 88; *Ann. and Mag. Nat. Hist.*, 1873, xii. p. 162.
Cervus antisiensis, *Gray, Ann. & Mag. Nat. Hist.*, 1873, xi. p. 310.

1375 *a.* Skull; male, with rudimentary anomalous horns. (Pl. xxxv. f. 1).
S. America. 58, 5, 4, 21. Zool. Society.

1375 *b.* Skull; female. With a small shallow pit in front of the orbit; intermaxillaries narrow above, not reaching quite to the nasal bones, without any prominence on the side of the head behind the orbits. The forehead between the eyes is strongly keeled. (Pl. xxxv. f. 2).
Peru. 73, 6, 27, 3.

15. XENELAPHUS, *Gray, Cat. Rum. Mam.* p. 88; *Ann. & Mag. Nat. Hist.*, 1873, xi. p. 220.

1. XENELAPHUS CHILENSIS, *Gray, Ann. & Mag. Nat. Hist.*, 1873, xii. p. 161.
Cervus chilensis, *Gay and Gervais, Ann. Sci. Nat.*, 1846, p. 21 (young); *Gay, Hist. d. Chili, Atlas,* tab. young animal and its skull.
Xenelaphus anomalocera, *Gray, Ann. & Mag. Nat. Hist.*, 1873, xi. p. 220.
Anomalocera huamel, *Gray, Scientific Opinion,* 1869, p. 385.
Xenelaphus huamel, *Gray, P. Z. S.,* 1869, p. 497, fig. 1 (horns) and p. 498, fig. 2 (skull of female).
Anomalocera leucotis, *Philippi in Wiegm. Arch.*, 1870, p. 46.
Xenelaphus leucotis, *Gray, Cat. Rum. Mam.* p. 89 (not part of synonyma).

1525 *a.* Animal, stuffed; adult male.
Xenelaphus huamel, *Gray, P. Z. S.,* 1869, p. 497, fig. (horns).
Peruvian Andes, Tinta. 69, 10, 15, 1. Whitely, jun.

1525 *b.* Animal, stuffed; adult female.
1525 *a.* Skull of "*b.*"
Xenelaphus huamel, *Gray, P. Z. S.,* 1869, p. 498 fig. (skull).
Peru, Tinta. 69, 10, 15, 2. Whitely, jun.

1525 *d.* Animal, stuffed; young.
1525 *d.* Skull of "*d.*"
Tinta. 69, 10, 15, 3. Whitely, jun.

1525 *c.* Animal, stuffed.
1525 *c.* Skull of "*c.*"
Tinta. 69, 10, 15, 4. Whitely, jun.

1525 *b.* Skull.
Tinta. 69, 10, 15, 5. Whitely, jun.

16. HUAMELA, *Gray, Ann. & Mag. Nat. Hist.* 1873, xi. p. 217; and 1873, xii. p. 162. Horns erect, simple, with a well-developed sub-basal anterior snag, and a subcentral snag. Skull destitute

of canine teeth in both sexes, with a large very deep sub-triangular tear-pit. The upper outer edge of the orbit thickened and produced behind into a conical prominence on the side of the forehead. The intermaxillary bones broad behind and reaching up to the nasals.

1. HUAMELA LEUCOTIS, *Gray*, *Ann. & Mag. Nat. Hist.*, 1873, xi. p. 219, fig. (skull and horns) ; 1873, xii. p. 162.
" Hoofed animal," *Hawkesworth's Voyages*, vol. i. p. 388.
Equus bisulcus, *Molina's Chili*, p. 320, 1782 (from Hawkesworth and other copiers of Molina).
Capreolus leucotis, *Gray*, *P. Z. S.*, 1849, p. 64, t. xii. (female) ; *Ann. & Mag. Nat. Hist.*, 1872, x. p. 445.
Capreolus? huamula, *Knowsley Menag*.
Furcifer huamel, *Gray*, *P. Z. S*., 1850, p. 236.
Xenelaphus leucotis (part), *Gray*, *Cat. Rum. Mam.* p. 89 (part of synonyma).

1584 *a*. Skin, in a bad state ; male.
1584 *a*. Skull of " *a*."
Huamela leucotis, *Gray*, *Ann. & Mag. Nat. Hist.*, 1873, xi. p. 218, fig. (skull).
W. Coast of Patagonia.　72, 10, 26, 1.
　　　　　　　Presented by Don Enrique Simpson.

1584 *b*. Skin, in a bad state ; female.
1584 *b*. Skull of " *b*"; without the lower jaw.
W. Coast of Patagonia.　72, 10, 26, 2.
　　　　　　　Presented by Don Enrique Simpson.

1584 *c*. Furriers' skin.
Capreolus leucotis (part), *Gray*, *Cat. Rum. Mam.* p. 227.
Xenelaphus leucotis (part), *Cat. Mam.* p. 89.
Patagonia.　50, 8, 2, 1.　Earl of Derby.

17. COASSUS, *Gray*, *Cat. Rum. Mam.* p. 91.

1. COASSUS NEMORIVAGUS, *Gray*, *Cat. Rum. Mam.* p. 91.

141 *c*. Animal, stuffed ; young.
S. America.

141 *a*. Animal, stuffed ; young.
Bolivia.　46, 7, 28, 39.　Bridges.

141 *b*. Skin ; adult male.
Plains near Santa Cruz.　47, 11, 22, 21.

685 *a*. Skull ; female.
S. America.　46, 2, 13, 3.

685 *b*. Skull.
S. America.

2. COASSUS SIMPLICICORNIS, *Gray*, *Cat. Rum. Mam.* p. 91.

1037 *a*. Animal, stuffed ; male.
British Guiana.　44, 9, 11, 107.　Schomburgh.

1037 *b*. Animal, stuffed ; female.
S. America.

1037 *c*. Animal, stuffed; male.
Brazils.

1037 *d*. Animal, stuffed; female.
Brazils. 41, 593.

1037 *e*. Skin.
S. America. 58, 6, 18, 7. Zool. Society.

1037 *a*. Skull; male.
S. America. 52, 12, 9, 6.

1037 *b*. Skull; young.
Surinam. 59, 9, 6, 106. Günther.

1037 *c*. Skull; very young.
S. America.

3. COASSUS RUFUS, *Gray, Cat. Rum. Mam.* p. 92 ; *Cat. Mam. B. M.,*
Ungulata, t. xxxv. f. 4 (skull).

684 *a*. Animal, stuffed; male.
Brazils. 41, 595.

684 *b*. Animal, stuffed; adult female.
Brazils. 41, 594.

684 *c*. Animal, stuffed; young.
St. Catharine's. 41, 6, 1, 26.

684 *d*. Skin; male.
St. Catharine's. 46, 6, 1, 23.

684 *e*. Skin; young.
St. Catharine's. 46, 6, 1, 25.

684 *f*. Skin; adult female.
St. Catharine's. 46, 6, 1, 24.

684 *g*. Skin; male; in a bad state.
Upper Ucayle. 66, 3, 28, 14.

684 *d*. Horns, on frontal bone.
S. America. Presented by Manuel Velez, Esq.

684 *a*. Skeleton; male.
S. America.

684 *c*. Skeleton; young male.
S. America. 46, 4, 21, 7.

684 *b*. Skull; female.
Para. 45, 8, 5, 9. Presented by R. Graham, Esq.

684 *c*. Skull.
S. America. 55, 6, 15, 7. Zool. Society.

4. COASSUS SUPERCILIARIS, *Gray, Cat. Rum. Mam.* p. 92.

a. Animal, stuffed; young male.
S. America. 49, 1, 12, 35.

b. Animal, stuffed; young.
b. Skull of "*b*."
S. America. 49, 9, 3, 1.

Y

c. Animal, stuffed.
 C. rufinus, *Puchéran, Arch. d. Mus.*, vi. p. 491.?
 S. America. 61, 12, 27, 13.

 5. COASSUS ——?

a. Skull of female, with small canine in the upper jaw. Nose slender.
 S. America. Jeude. 67, 4, **12**, 248.

 6. COASSUS ——?

Homelophus inornatus, *Gray, Cat. Rum. Mam.* p. 90.
a. Animal, stuffed; male.
 Hab. ——? 51, 8, 29, 7. Zool. Society.

 7. COASSUS? (Cervus) WHITELYI, *Gray, Ann. & Mag. Nat. Hist.*,
  1873, xii. p. 163.

1618 *a.* Skull of a rather young animal, with only five grinders on
  each side, which yet appear to be fully formed, and is unlike
  the skull of any South-American deer in the Museum Collection,
  the brain-cavity being much larger and more ventricose compared with the compressed face than in any other known skull;
  and it has rudimentary canines, which are not to be observed in
  any species of *Coassus* or smaller South-American deer. The
  skull is 6¾ inches long, and 3½ inches wide in the lower edge of
  the middle of the orbital opening (which is the widest part of
  the skull), and 3⅔ inches from the end of the occiput to the front
  of the orbit, and 3½ inches from the front of the orbit to the end
  of the intermaxillaries. There is a rather elongate groove over
  each orbit, as in the skull of *Coassus nemorivagus;* but the
  brain-case of this skull is very much narrower, and has a keel in
  the centre of the forehead, which is entirely absent in the flat
  broad forehead of *Cervus Whitelyi*. There is a moderately deep,
  concave, rounded pit for the tear-gland, and two perforations
  for the passage of vessels through the orbit, just behind the
  lachrymal pit. The brain-case is oblong, narrowed above, at
  the upper edge of the orbits. At the lower edge of the orbits it
  is much expanded out, being the widest part of the skull. The
  face, from the upper edge of the orbits, is gradually, and from
  the lower edge of the orbits rapidly, attenuated as far as the
  front end of the grinders. The nose, from the front end of the
  grinders, slender, compressed, with the front half of its length
  rather narrowed on the sides. The nasal bones moderate, the
  middle of the hinder end being broadly produced between the
  front of the frontals, which I have not observed in any other
  deer. The intermaxillary bones very slender in front, the hinder
  half becoming much broader above, and attached to the sides of
  the front of the nasals—more so than in any South-American
  deer that I have yet observed. (Pl. xxxii. f. 2).
Peru, Conipata. 73, 6, 27, 2. Whitely.

18. PUDU, *Gray, Cat. Rum. Mam.* p. 92.

    1. PUDU HUMILIS, *Gray, Cat. Rum. Mam.* p. 92.

972 *a.* Animal, stuffed; female.
Cervus humilis, *Bennett, P. Z. S.*, 1831, p. 27.
Chili. 55, 12, 24, 284. Zool. Society.

972 *b.* Animal, stuffed; female.
972 *a.* Skull of "*b.*"
Pudu chilensis, *Gray, Cat. Mam. B. M., Ungul.*, t. xxxvi. f. 1.
Chili. 50, 11, 29, 6. Zool. Society.

972 *c.* Animal, stuffed; female.
Chili. 54, 12, 6, 6. Zool. Society.

Family 21. CERVULIDÆ, *Gray, Cat. Rum. Mam.* p. 93.

1. CERVULUS, *Gray, Cat. Rum. Mam.* p. 93.

    1. CERVULUS MOSCHATUS, *Gray, Cat. Rum. Mam.* p. 93; *Cat. Mam. B. M., Ungulata*, t. xxxii. f. 2 (skull and horn).

701 *a.* Animal, stuffed; young male.
India. 39, 12, 20, 2.

701 *b.* Animal, stuffed; adult female.
India. 42, 4, 13, 2.   Presented by B. H. Hodgson, Esq.

701 *c.* Animal, stuffed; adult female.
India.

701 *d.* Animal, stuffed; female, three parts grown.
701 *o.* Skull of "*d,*" very imperfect.
Nepal. 45, 4, 10, 6.   Presented by B. H. Hodgson, Esq.

701 *e.* Animal, stuffed; very young, spotted.
701 *bb.* Skull of "*e.*"
Nepal. 58, 6, 24, 15.   Presented by B. H. Hodgson, Esq.

701 *f.* Skin; adult male.
Nepal. 43, 1, 12, 123.   Presented by B. H. Hodgson, Esq.

701 *g.* Skin; adult female.
Nepal. 58, 6, 24, 14.   Presented by B. H. Hodgson, Esq.

701 *h.* Skin; male.
Nepal. 55, 1, 20, 6.
              Presented by His Highness Maharajah Dhuleep Singh.

701 *i.* Skin; young female.
Nepal. 58, 6, 24, 13.   Presented by B. H. Hodgson, Esq.

701 *j.* Skin, without feet.
China, Hainan. 70, 2, 10, 26. Swinhoe.

701 *k.* Skin, without head and feet.
China, Hainan. 70, 2, 10, 26. Swinhoe.

701 *l.* Skin, without head and feet.
India.   Presented by Gen. Hardwicke.

701 *m*. Skin; young male.
    Nepal. 58, 6, 24, 11.   Presented by B. H. Hodgson, Esq.

701 *f*. Horns, on frontal bone.
    India.

701 *j*. Horns, on frontal bone.
    Nepal. 45, 1, 8, 193.   Presented by B. H. Hodgson, Esq.

701 *k*. Horns, on frontal bone.
    Nepal. 45, 1, 8, 172.   Presented by B. H. Hodgson, Esq.

701 *t—v*. Horns, on frontal bone.
    Nepal. 58, 6, 24, 182—184.   Presented by B. H. Hodgson, Esq.

701 *w*, *x*. Horns, on frontal bone.
    India. 56, 5, 6, 66.   Presented by Prof. Oldham.

701 c^2, d^2. Horns, on frontal bone.
    China, Hainan. 70, 2, 10, 82—3.   Swinhoe.

701 *a*. Skull; male.
    Northern Circars. 38, 3, 10, 43.

701 *c*. Skull; male.
    Northern Circars. 43, 1, 26, 13.

701 *d*. Skull; male.
    Nepal. 45, 1, 8, 189.   Presented by B. H. Hodgson, Esq.

701 *e*. Skull; male.
    India.

701 *g*. Skull; male.
    Nepal. 45, 1, 8, 190.   Presented by B. H. Hodgson, Esq.

701 *h*. Skull; male.
    Nepal. 45, 1, 8, 189.   Presented by B. H. Hodgson, Esq.

701 *i*. Skull; female.
    Nepal. 45, 1, 8, 191.   Presented by B. H. Hodgson, Esq.

701 *m*. Skull; young male.
    Nepal. 48, 6, 11, 24.   Presented by B. H. Hodgson, Esq.

701 *n*. Skull; young female.
    India.

701 *p*. Skull; young female.
    Darjeeling. 56, 5, 6, 63.   Presented by Prof. Oldham.

701 *q*. Skull; adult female.
    Darjeeling. 56, 5, 6, 64.   Presented by Prof. Oldham.

701 *r*. Skull; adult male.
    Nepal.   Presented by B. H. Hodgson, Esq.

701 *s*. Skull; male.
    Nepal.   Presented by B. H. Hodgson, Esq.

701 a^2. Skull; male.
    India.

701 b^2. Skull; young.
    Nepal.   Presented by B. H. Hodgson, Esq.

2. **Cervulus** curvostylis, *Gray, Cat. Rum. Mam.* p. 94.

1619 *a*. Skull, without the lower jaw.
Siam, Pachebon. 61, 6, 1, 8. Mouhot.

3. **Cervulus** tamulicus, *Gray, Cat. Rum. Mam.* p. 94.

701 *b*. Skull; male.
Deccan. Presented by Col. Sykes.

4. **Cervulus** Reevesii, *Gray, Cat. Rum. Mam.* p. 94.

1524 *a*. Animal, stuffed; female.
China. 50, 11, 22, 12. Zool. Society.

1524 *b*. Animal, stuffed; female.
Cervus Reevesii, *Ogilby, P. Z. S.* 1838, p. 105.
China. 53, 8, 29, 49. Russell Reeves.

1524 *c*. Animal, stuffed; male.
Cervus Reevesii, *Ogilby, P. Z. S.* 1838, p. 105.
China. 55, 2, 24, 283. Russell Reeves.

1524 *d*. Animal, stuffed; very young, without spots.
Amoy. 60, 1. Swinhoe.

1524 *e*. Skin; young female.
1524 *c*. Skull of " *e*."
China, Ningpo. 72, 9, 3, 2. Swinhoe.

1524 *f*. Skin.
Formosa. 62, 12, 24, 3. Swinhoe.

1524 *a*. Skeleton; male.
C. Reevesii, *Gray, Cat. Rum. Mam.* t. 2, f. 3 (skull and horns).
China. 62, 12, 26, 2. Zool. Society.

1524 *b*. Skeleton; male.
China. 72, 9, 3, 8. Swinhoe.

5. **Cervulus** Sclateri, *Swinhoe, P. Z. S.* 1872, p. 814.

1620 *a*. Animal; adult. Fur reddish brown, minutely punctulated
with yellowish grey; throat rather paler; chin, inside of ears,
and end of tail, white; orbits, and a streak on each side of the
forehead to the base of the horns, black.
1620 *a*. Skull of " *a*." Nose broad; same length as head from front
of orbit.
Cervulus Sclateri, *Swinhoe, P. Z. S.* 1872, p. 814.
China, Ningpo. 72, 9, 3, 1. Swinhoe.

1620 *c*. Animal; young. Rather paler, with a row of spots on each
side of dorsal line, and two oblique rows of spots on each side,
white; orbits, streak over each eye, middle of the crown, and
central nuchal streak, black; face and throat underneath paler
brown.
1620 *b*. Skull and bones of body of " *c*." Nose very short, tapering.
Cervulus Sclateri, *Swinhoe, P. Z. S.* 1872, p. 814.
China, Ningpo. 72, 9, 3, 2. Swinhoe.

Family 22. MOSCHIDÆ, *Gray, Cat. Rum. Mam.* p. 95.

1. HYDROPOTES, *Gray, Cat. Rum. Mam.* p. 95; *Brooke, P. Z. S.*
    1872, p. 524, figs. 1, 2 (skull).

  1. HYDROPOTES INERMIS, *Gray, Cat. Rum. Mam.* p. 95.

1551 *a*. Animal, stuffed.
1551 *a*. Skull of " *a*."
    Hydropotes inermis, *Swinhoe, P. Z. S.* 1870, p. 90, t. vi. (animal),
      t. vii. (skull).
    Shanghai. 70, 7, 18, 15. Swinhoe.

1551 *b*. Skin ; male.
1551 *c*. Skull of " *b*."
    Shanghai. 72, 9, 3, 5. Swinhoe.

1551 *c*. Skin ; male.
1551 *b*. Skull of " *c*."
    Chin-kiang. 72, 9, 3, 4. Swinhoe.

1551 *e*. Skeleton ; male.
    Shanghai. 72, 9, 3. Swinhoe.

1551 *d*. Skull ; female.
    Shanghai. 72, 9, 3, 6. Swinhoe.

2. MOSCHUS, *Gray, Cat. Rum. Mam.* p. 96.

  1. MOSCHUS MOSCHIFERUS, *Gray, Cat. Rum. Mam.* p. 96.

676 *a*. Animal, stuffed.   .
    M. moschiferus.
    Java.   Hon. E. India Company.

676 *b*. Animal, stuffed.
    M. leucogaster, *Hodgson.*
    Nepal. 43, 1, 12, 95.   Presented by B. H. Hodgson, Esq.

676 *c*. Animal, stuffed.
    M. leucogaster, *Hodgson.*
    Nepal. 55, 1, 20, 9.   Pres. by H. H. Maharajah Dhuleep Singh.

676 *d*. Animal, stuffed ; young.
    M. leucogaster, *Hodgson.*
    Nepal. 45, 1, 8, 329.   Presented by B. H. Hodgson, Esq.

678 *e*. Animal, stuffed ; male.
    M. chrysogaster, *Hodgson.*
    Nepal. 43, 1, 12, 93.   Presented by B. H. Hodgson, Esq.

677 *f*. Animal, stuffed.
677 *a*. Skull of " *f*."
    M. sibiricus, *Pallas.*
    Siberia. 42, 4, 29, 75.

677 *g*. Animal, stuffed.
677 *b*. Skull of " *g*."
    M. sibiricus, *Pallas.*
    Siberia. 42, 4, 29, 76.

676 *d*. Skeleton, very imperfect.
Nepal. 45, 1, 12, 448.　　　　　Presented by B. H. Hodgson, Esq.

676 *e*. Skeleton ; feet wanting.
Nepal. 45, 1, 22, 449.　　　　　Presented by B. H. Hodgson, Esq.

676 *c*. Bones of the body.
Nepal.　　　　　　　　　　　Presented by B. H. Hodgson, Esq.

678 *a*. Skull ; male.
Moschus chrysogaster, *Gray, Cat. Mam. B. M., Ungulata,* t. xxv.
f. 1 (skull).
Nepal.　　　　　　　　　　　Presented by B. H. Hodgson, Esq.

678 *b*. Skull.
Moschus chrysogaster, *Hodgson.*
Nepal. 43, 1, 12, 94.　　　　　Presented by B. H. Hodgson, Esq.

676 *a*. Skull ; female.
Nepal. 45, 1, 8, 357.　　　　　Presented by B. H. Hodgson, Esq.

676 *b*. Skull.
Nepal. 43, 1, 26, 4.　　　　　Presented by B. H. Hodgson, Esq.

676 *f*. Skull ; male.
Nepal. 48, 6, 11, 26.　　　　　Presented by B. H. Hodgson, Esq.

676 *g*. Skull; male.
Nepal.　　　　　　　　　　　Presented by B. H. Hodgson, Esq.

676 *h*. Skull.
Karhm.　　　　　　　　　　　Presented by Prof. Oldham.

676 *i*. Skull ; male.
Nepal. 43, 1, 12, 97.　　　　　Presented by B. H. Hodgson, Esq.

676 *j*. Skull ; male.
Nepal. 43, 1, 12, 98.　　　　　Presented by B. H. Hodgson, Esq.

676 *k*. Skull ; female.
India.

Family 23. TRAGULIDÆ, *Gray, Cat. Rum. Mam.* p. 97.

1. MEMINNA, *Gray, Cat. Rum. Mam.* p. 97.

　　1. MEMINNA INDICA, *Gray, Cat. Rum. Mam.* p. 97 ; *Cat. Mam. B. M.,
　　Ungulata,* t. xxiv. f. 3 (skull).

679 *a*. Animal, stuffed ; female.
India. 41, 837.

679 *b*. Animal, stuffed ; female.
India. 47, 4, 30, 3.

679 *c*. Animal, stuffed ; female.
679 *a*. Skull of " *c*."
Meminna indica, *Gray, Cat. Mam. B. M., Ungulata,* t. xxiv. fig. 3
(skull).
India. 45, 8, 7, 9.

679 *d*. Animal, stuffed ; young.
India.　　　　　　　　　　　Presented by W. Elliot, Esq.

679 *e*. Animal, stuffed ; very young.
India. 38, 3, 13, 47.

679 *f*. Animal, stuffed.
M. Malaccensis, *Gray, Cat.*
Singapore. 42, 5, 26, 19.

679 *b*. Bones of the body.
India.

679 *c*. Skeleton.
India. Zool. Society.

2. TRAGULUS, *Gray, Cat. Rum. Mam.* p. 97.

1. TRAGULUS STANLEYANUS, *Gray, Cat. Rum. Mam.* p. 98.

827 *a*. Animal, stuffed.
Ceylon. 53, 8, 29, 38. Zool. Society.

827 *b*. Animal, stuffed.
India. 48, 10, 11, 6.

827 *c*. Animal, stuffed ; very young.
Born in the Zoological Gardens. 53, 8, 29, 40.

827 *b*. Skull.
Tragulus Stanleyanus, *Gray, Cat. Mam. B. M., Ungulata*, t. xxv. f. 3.
Pachebore. 61, 4, 12, 20.

2. TRAGULUS JAVANICUS, *Gray, Cat. Rum. Mam.* p. 98.

52 *b*. Animal, stuffed ; adult.
Sumatra.   Presented by Lady Raffles.

52 *c*. Animal, stuffed ; male.
Java.   Hon. E. India Company.

52 *a*. Skin.
Java.   Presented by Gen. Hardwicke.

52 *d*. Skin.
Java ?   40, 3, 6, 53.

1361 *a*. Skeleton.
Java ?   52, 6, 26, 4.

3. TRAGULUS FULVIVENTER, *Gray, Cat. Rum. Mam.* p. 98.

853 *a*. Animal, stuffed ; female.
T. Kanchil, *Raffles.*
Java.   Hon. E. India Company.

853 *b*. Animal, stuffed ; female.
T. Kanchil.
Java ?   38, 7, 13, 2.

853 *c*. Animal, stuffed.
T. Kanchil.
Java.   Hardwicke bequest.

853 *d*. Animal, stuffed.
T. Kanchil, *Horsfield*.
India.   Presented by E. Burton, Esq.

853 *e*. Animal, stuffed.
T. Kanchil, *Horsfield*.
Java. 47, 4, 30, 9.

853 *f*. Animal, stuffed; very young.
T. Kanchil, *Horsfield*.
Java.   Hardwicke bequest.

853 *g*. Animal, stuffed; very young.
Singapore. 42, 4, 12, 14.

853 *h*. Animal, stuffed; very young.
Born in the Zoological Gardens. 53, 8, 29, 39.

853 *i*. Animal, stuffed; female.
T. affinis, *Gray, P. Z. S.*, 1861, p. 138.
Malacca. 38, 8, 1, 13.

853 *j*. Animal, stuffed; male.
T. affinis, *Gray*.
Camboja. 61, 4, 12, 7. Mouhot.

853 *k*. Animal, stuffed; young.
T. affinis, *Gray*.
Camboja. 61, 4, 12, 6. Mouhot.

853 *f*. Skeleton; female.
India. 60, 3, 18, 29. Zool. Society.

853 *g*. Skeleton, with a bony shield over the pelvis.
India? 68, 12, 29, 36. Zool. Society.

853 *a—c*. Skulls, imperfect.
India? 48, 11, 51—3.

853 *d*. Skull.
India? Zool. Society.

853 *e*. Skull.
India. 56, 5, 6, 67.   Presented by Professor Oldham.

4. TRAGULUS ———— ?

a—d. Four skulls; male, female, and young.
Hab. ———— ? 67, 4, 12, 293—4, 6, 7. Lidth de Jeude.

Family 24. HYEMOSCHIDÆ, *Gray, Cat. Rum. Mam.* p. 99.

1. HYEMOSCHUS, *Gray, Cat. Rum. Mam.* p. 99.

1. HYEMOSCHUS AQUATICUS, *Gray, Cat. Rum. Mam.* p. 99; *Cat. Mam. B. M., Ungulata*, t. xxv. f. 2 (skull).

z

680 *a.* Animal, stuffed; adult.
680 *b.* Skull of "*a.*"
  Hyem. aquaticus, *Gray, Cat. Rum. Mam. B. M., Ungulata,* t. xxv.
    f. 3 (skull).
  Sierra Leone. 44, 9, 7, 1.   Presented by the Earl of Derby.

680 *b.* Animal, stuffed.
680 *a.* Skull of "*b.*"
  Hyem. aquaticus, *Gray, Ann. & Mag. Nat. Hist.* 1845, xvi. p. 350.
  Sierra Leone. 44, 3, 22, 1.   Presented by the Earl of Derby.

680 *c.* Animal, stuffed; young.
  Gambia. 46, 10, 2, 3.   Presented by the Earl of Derby.

680 *c.* Skeleton, imperfect.
  Sierra Leone. Whitfield.

680 *e.* Skeleton, imperfect.
  Gambia. Zool. Society.

680 *d.* Skull, and limbs.
  Gambia.   Presented by the Earl of Derby.

680 *f.* Skull.
  Gambia. 58, 5, 4, 452. Zool. Society.

    Sub-order 5. TYLOPODA, *Gray, Cat. Rum. Mam.* p. 100.
    Family 25. CAMELIDÆ, *Gray, Cat. Rum. Mam.* p. 100.

1. CAMELUS, *Gray, Cat. Rum. Mam.* p. 100.

  1. CAMELUS ARABICUS, *Gray, Cat. Rum. Mam.* p. 100.

672 *a.* Animal, stuffed; white variety.
672 *b.* Skull of "*a.*"
  Arabia? 46, 10, 26, 41.

672 *b.* Animal, stuffed; young.
  Arabia? 47, 3, 27, 37.

672 *c.* Skin; white variety.

   ————

672 *a.* Skull.
  Camelus arabicus, *Gray, Cat. Mam. B. M., Ung.* tab. **xxiii.** fig. 3.
   ———— Mantell.

672 *e.* Skeleton.
  N. Africa.

  2. CAMELUS BACTRIANUS, *Gray, Cat. Rum. Mam.* p. 100.

673. Skeleton.
  Persia. Warwick.

  2. LLAMA, *Gray, Cat. Rum. Mam.* p. 101.

  1. LLAMA VICUGNA, *Gray, Cat. Rum. Mam.* p. 101.

675 *a.* Animal, stuffed; male.
  Bolivia. 46, 10, 16, 16. Bridges.

675 *b*. Animal, stuffed ; female.
Bolivia ?

675 *b*. Skull.
Bolivia. 46, 11, 20, 4. Bridges.

675 *c*. Skeleton ; male.
Bolivia ? 61, 1, 18, 3. Zool. Society.

675 *a*. Skull ; male.
Ll. vicugna, *Gray, Cat. Mam. B. M., Ungul.*, tab. xxiv. fig. 2.
Bolivia. 46, 10, 16, 16. Bridges.

2. LLAMA PACOS, *Gray, Cat. Rum. Mam.* p. 101.

1360 *a*. Animal, stuffed ; adult, black.
S. America.   Presented by the Earl of Derby.

1360 *b*. Animal, stuffed ; three-parts grown, black.
S. America.

1360 *c*. Animal, stuffed ; adult, with very thick black wool.
1360 *a*. Skull of "*c*."
S. America. 49, 10, 2, 2.   Presented by W. Danston, Esq.

1360 *d*. Animal, stuffed ; young, white and brown.
S. America.

1360 *e*. Animal, stuffed ; just born, black.
Knowsley.   Presented by the Earl of Derby.

1360 *c*. Skeleton ; male.
S. America. Zool. Society.

1360 *e*. Skeleton ; female.
S. America. 67, 2, 24, 9. Zool. Society.

1360 *d*. Skull.
S. America. 55, 12, 26, 149. Zool. Society.

3. LLAMA GUANACUS, *Gray, Cat. Rum. Mam.* p. 101.

674 *a*. Animal, stuffed ; adult.
S. America.   Presented by Sir W. Burnett and Capt. Fitzroy.

674 *b*. Animal, stuffed ; adult.
S. America.   Presented by Charles Darwin, Esq.

674 *c*. Animal, stuffed ; half-grown.
S. America. 46, 1, 22, 4.

674 *d*. Animal, stuffed ; young.
S. America. 54, 5, 11, 2.

674 *e*. Skin ; young.
674 *e*. Skull of "*e*."
S. America. 55, 7, 3, 1. Zool. Society.

674 *a*. Skeleton.
Chili.

674 *b*. Skull.
Ll. guanacus, *Gray, Cat. Mam. B. M., Ung.* tab. xxiv. fig. 1.
Chili. 44, 10, 7, 34. Bridges.

674 *c.* Skull.
　　Chili?

674 *d.* Skull.
　　Chili.　46, 4, 8, 7.　Warwick.

　　4. LLAMA GLAMA, *Gray, Cat. Rum. Mam.* p. 101.

a. Animal; adult, white.
　　S. America.

ADDITIONS.

SCOPOPHORUS OUREBI, p. 90, *add :*—

Skins of male, female, and young male.
　　Abyssinia, Bogos Country.　Dombelas.　73, 8, 29, 4, 5, 6.

ALCELAPHUS, p. 114, *Synopsis of Species, Gray, Ann. & Mag. Nat. Hist.* 1873, xii. p, 342.　*Add :*—

1* ALCELAPHUS TORA, *Gray, ' Nature,'* 1873, Sept. 4, p. 364; *Ann. & Mag. Nat. Hist.* 1873, xii. p. 341.　Bright pale brown.　Rump, inside of ears, and hinder side of legs, whitish brown.　Horns expanded, rather recurved inwards at the tips; of the female smaller, more slender.

a. Skin of adult male.
　　Skull and horns of "*a*," Pl. xli.
　　Abyssinia, Bogos Country.　Dombelas.　73, 8, 29, 1.

b. Skin of female.
b. Skull of "*b*," Pl. xli. f. 2.
　　Abyssinia, Bogos Country.　Dombelas.　73, 8, 29, 2.

c. Skin of nearly adult male.　Horns shorter and thicker.　*Ann. & Mag. Nat. Hist.* 1873, xii. p. 341.
　　Abyssinia.　Dombelas.　73, 2, 24, 12.

d. Skeleton of adult male.
　　Abyssinia.　Dombelas.　73, 8, 29, 3.

INDEX

EXPLANATION OF THE PLATES.

* The nasal cavity of this genus is very complicated. It has a large maxillary palatine sinus, occupying the whole length and breadth of the palate, with a large opening on each side the vomer in front, just within the interior nostrils, separated from the upper nasal cavity by a regularly arched bony lamina. The pair of nasal cavities separated by a large well-developed bony

vomer.   Each of the cavities has an opening in the middle of the outer side, between the turbinal bones, into a narrow bony, sinuous tube on each side of the palatine cavity above described, which each has a small narrower opening in the front of the palate behind the cutting teeth and a larger one behind, forming the interior nasal opening.   The turbinals are attached to the upper outer surface of the upper nasal cavity, and have a large convoluted lamina on the lower side, and are met by the lower turbinals, the aperture to the nasal tube being between the turbinals.   The auricular bulla is cellular internally.   This structure is very peculiar, and very unlike the simple nasal cavity, direct from the anterior to the posterior nostrils, which is found in the Pig, *Phacocheir*, and in all the other *Pecora* and *Bellua* that I have examined or seen figured, and in them there is an aperture on each side in front of the palate, between the palatine and intermaxillary bones.

The section of *Dicotyles labiatus* is very similar.

E. NEWMAN, PRINTER, DEVONSHIRE STREET, BISHOPSGATE.

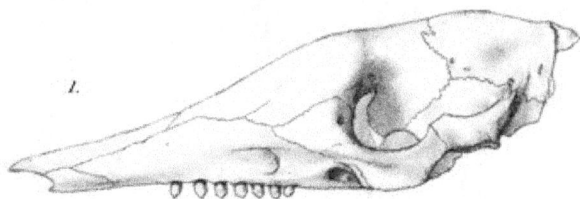

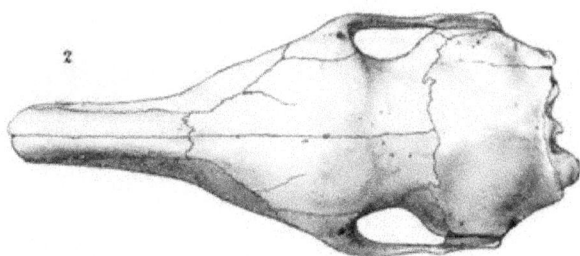

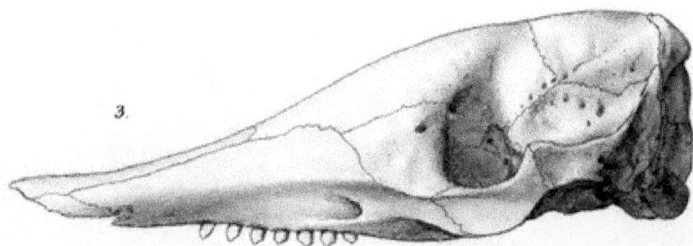

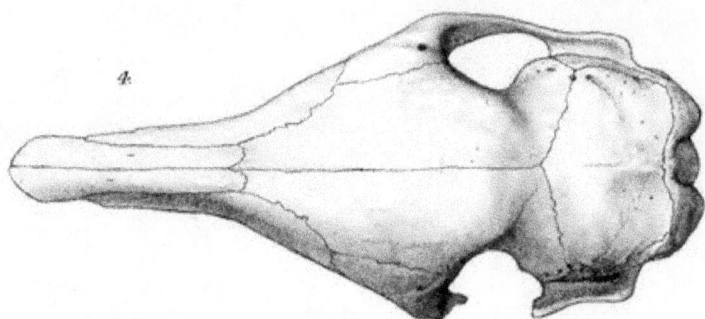

G. H. Ford & C. L. Griesbach.   Mintern Bros imp

1. 2. *Tatusia peba* p. 13.
3. 4. *Tatusia leptorhynchus* p 15.

1

2

3.

4.

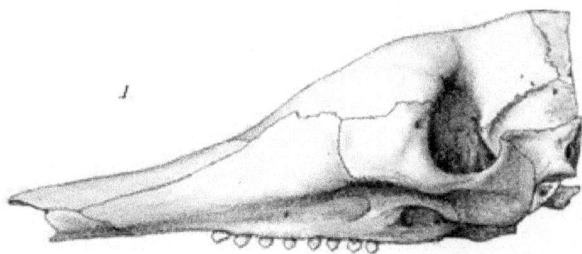

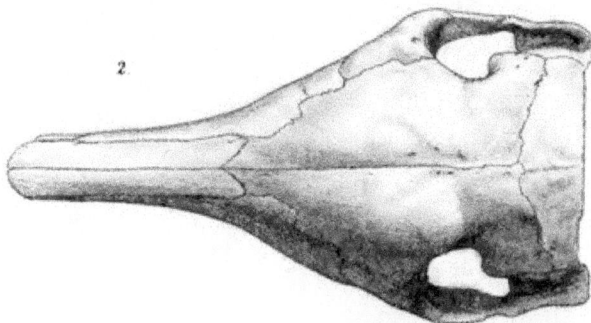

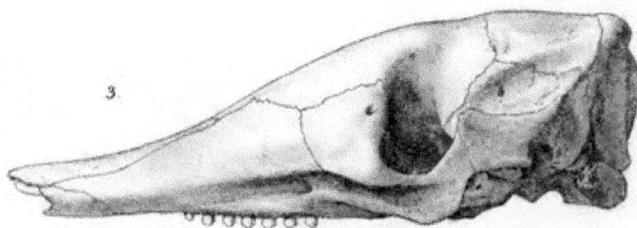

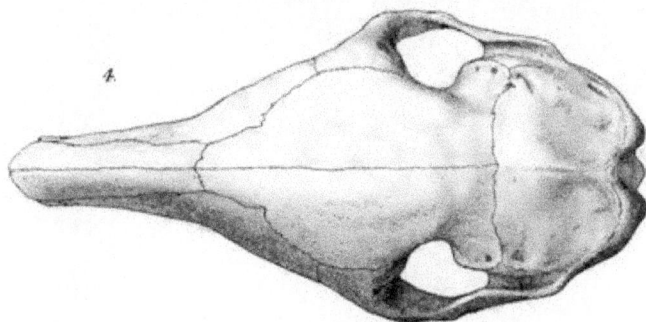

G.H.Ford & C.L. Griesbach.   Mintern Bros imp.

1. 2. Tatusia Granadiana p 14.
3. 4. Tatusia mexicana.

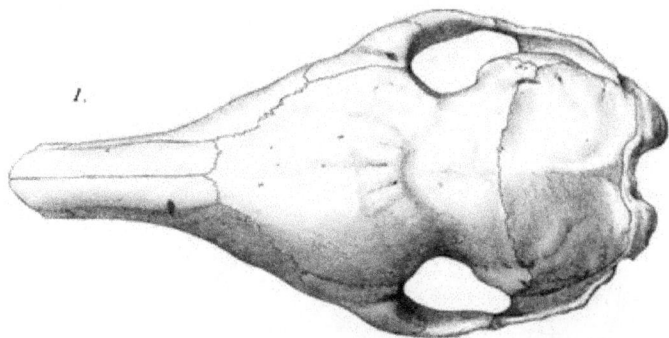

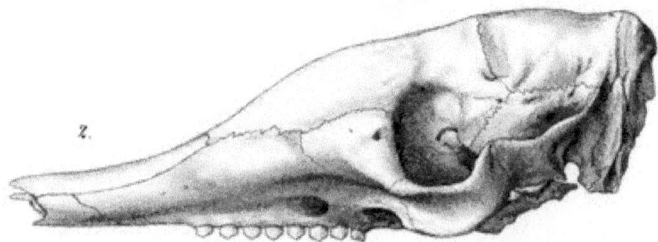

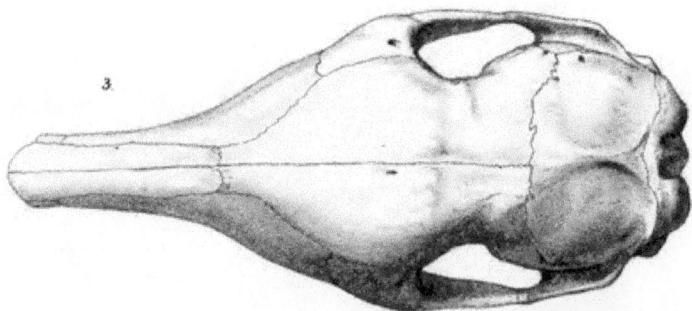

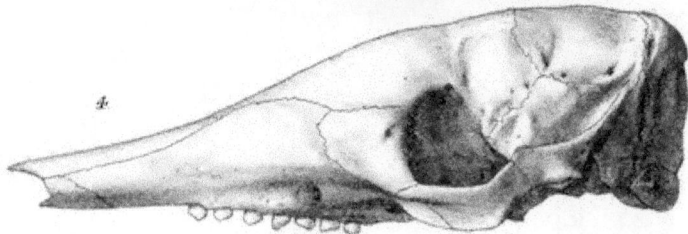

1. 2. *Tatusia   brevirostris,   p 15.*
3. 4. *Tatusia   boliviensis,   p. 16.*

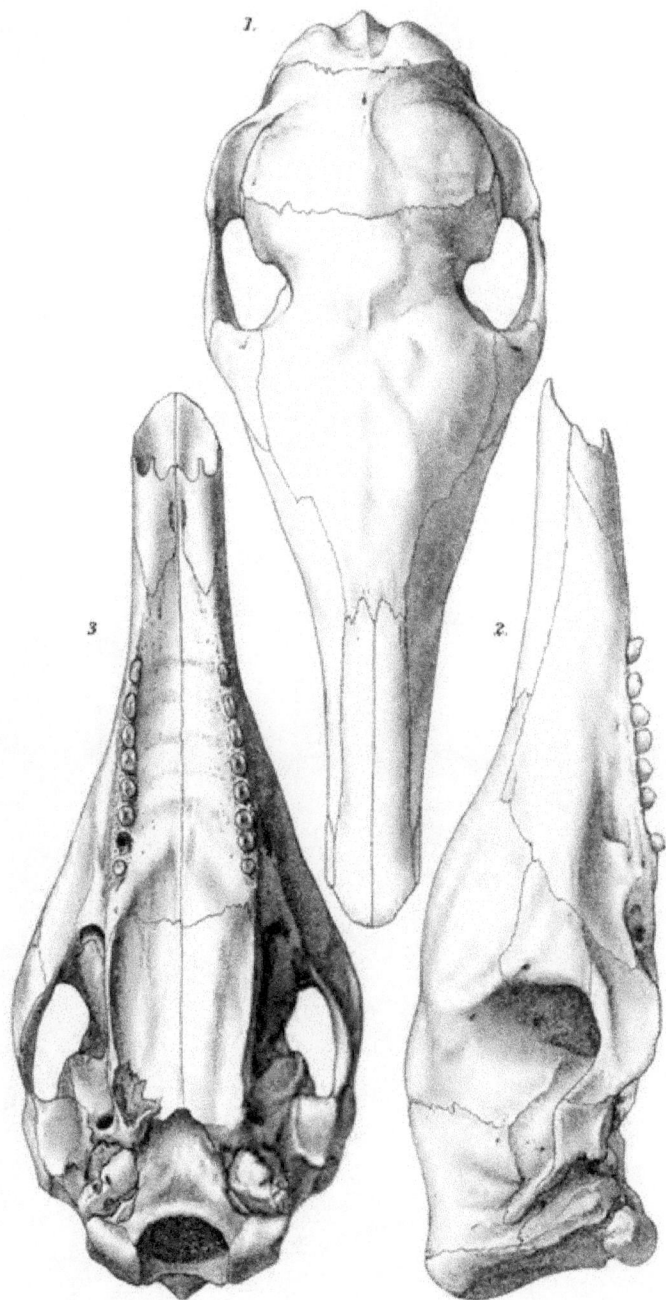

1. 2. 3. Praopus Kappleri p. 18.

1.

2.

1457 a

3.

731. C

4.

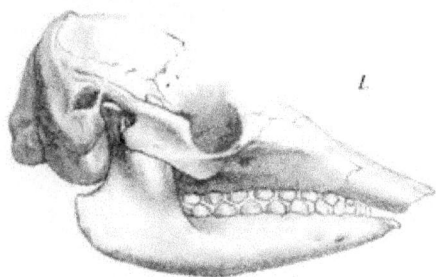

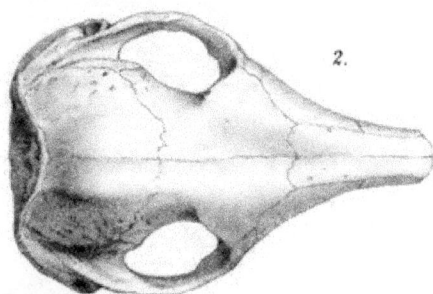

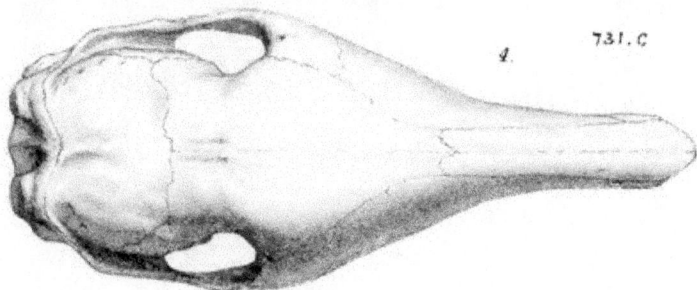

G.H.Ford & C.L.Griesbach.   · Mintern Bros imp.

1, 2. *Chætophractus vellerosus*, p. 19.
3, 4. *Tolusia leptocephala*, p. 16.

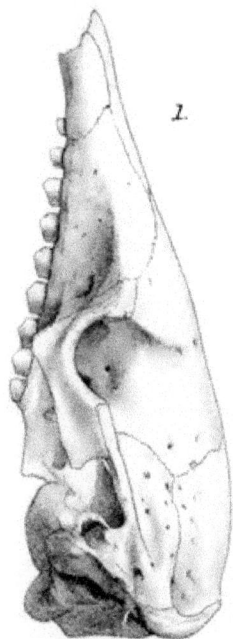

1.

914.6.

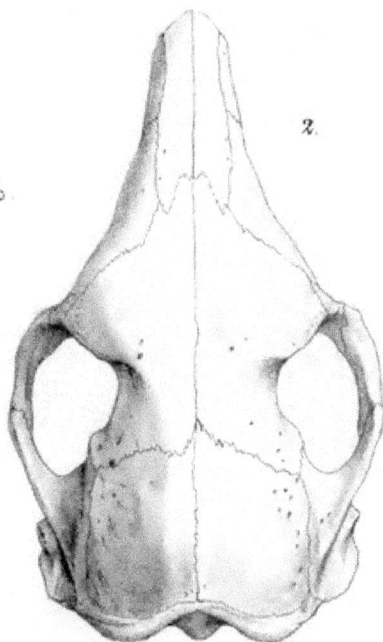

2.

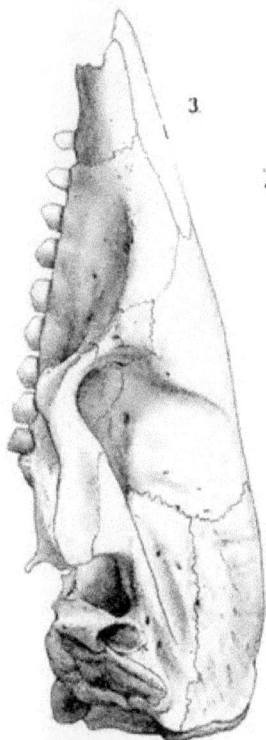

3.

720.a.

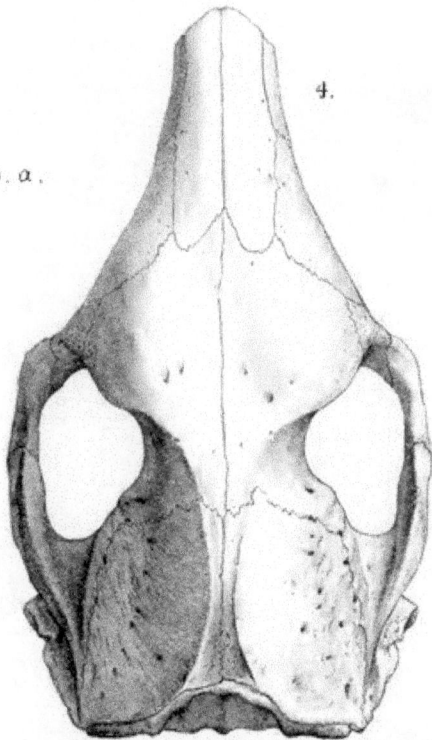

4.

G.H.Ford & C.L.Griesbach.

Mintern Bros. imp.

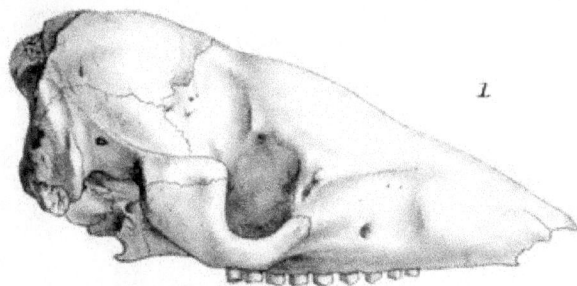

1.

51.8.25.9.

2.

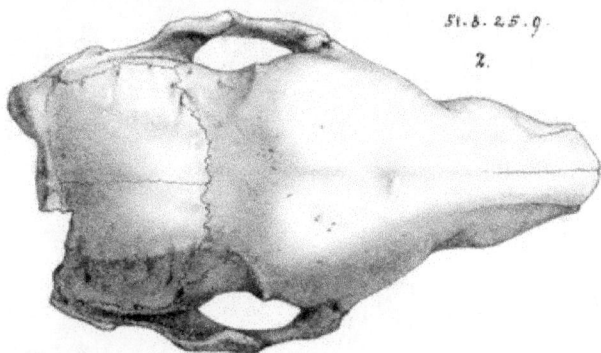

3.

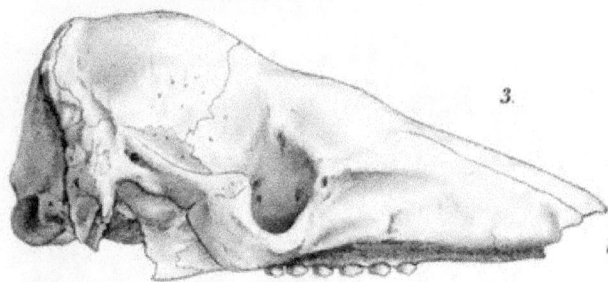

31.8.25.10.

4.

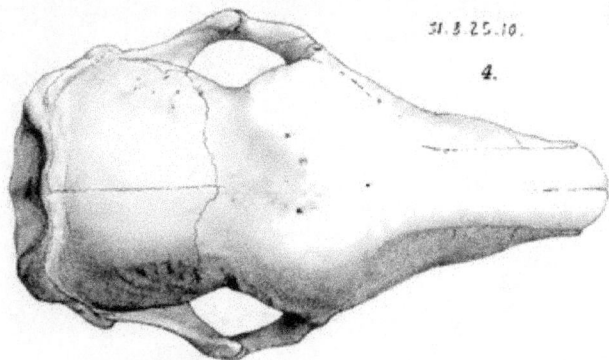

th...
Skull from
Demarara 124
Su. p. 23. he
unicinctus

G.H. Ford & C.L. Griesbach.   Mintern Bros imp.

1, 2. Zenurus latirostris, p. 22.
3, 4. Ziphila lugubris, p. 23.

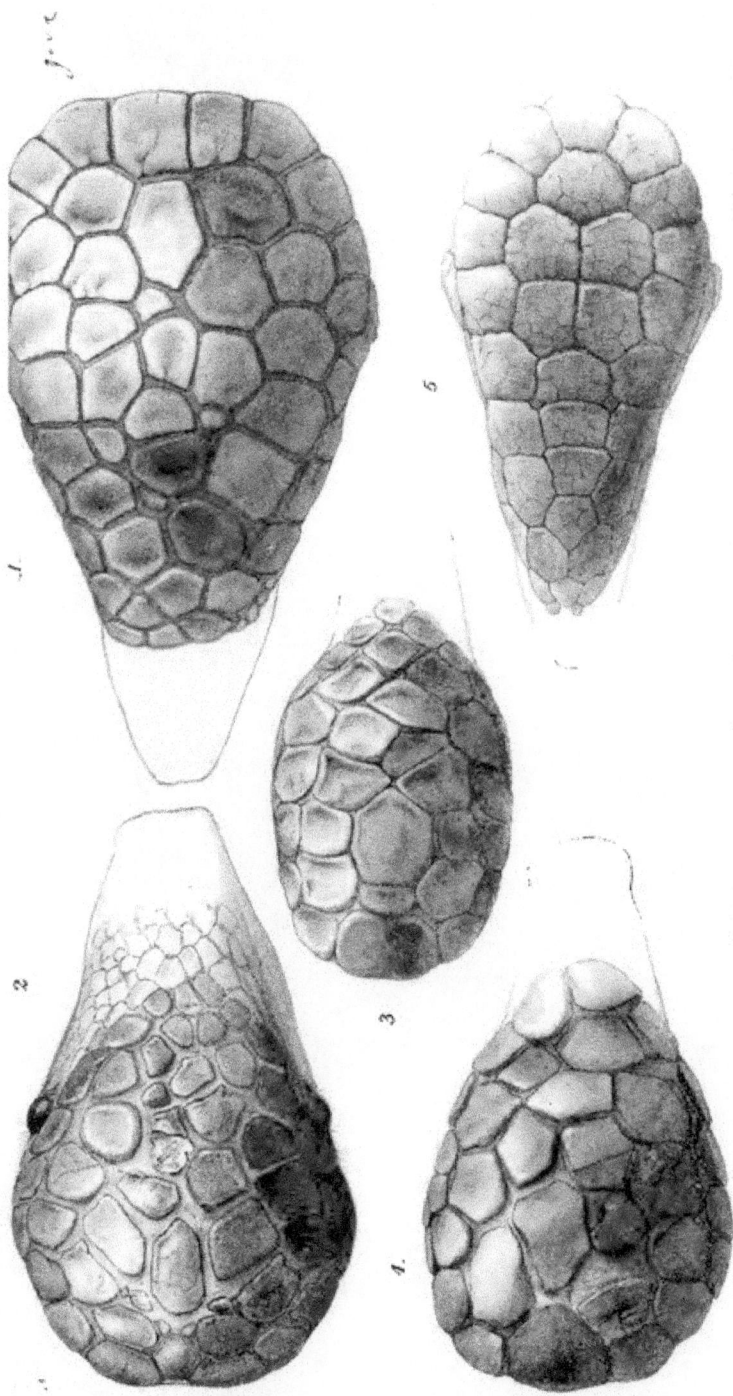

1. Xenurus *unicinctus*, p. 21.
2. " *latirostris*, p. 22.
3, 4. *Ziphila lugubris*, p. 23.
5. *Cheloniscus unicinctus*, p. 24.

G. H. Ford & C. L. Griesbach

Mintern Bros. imp.

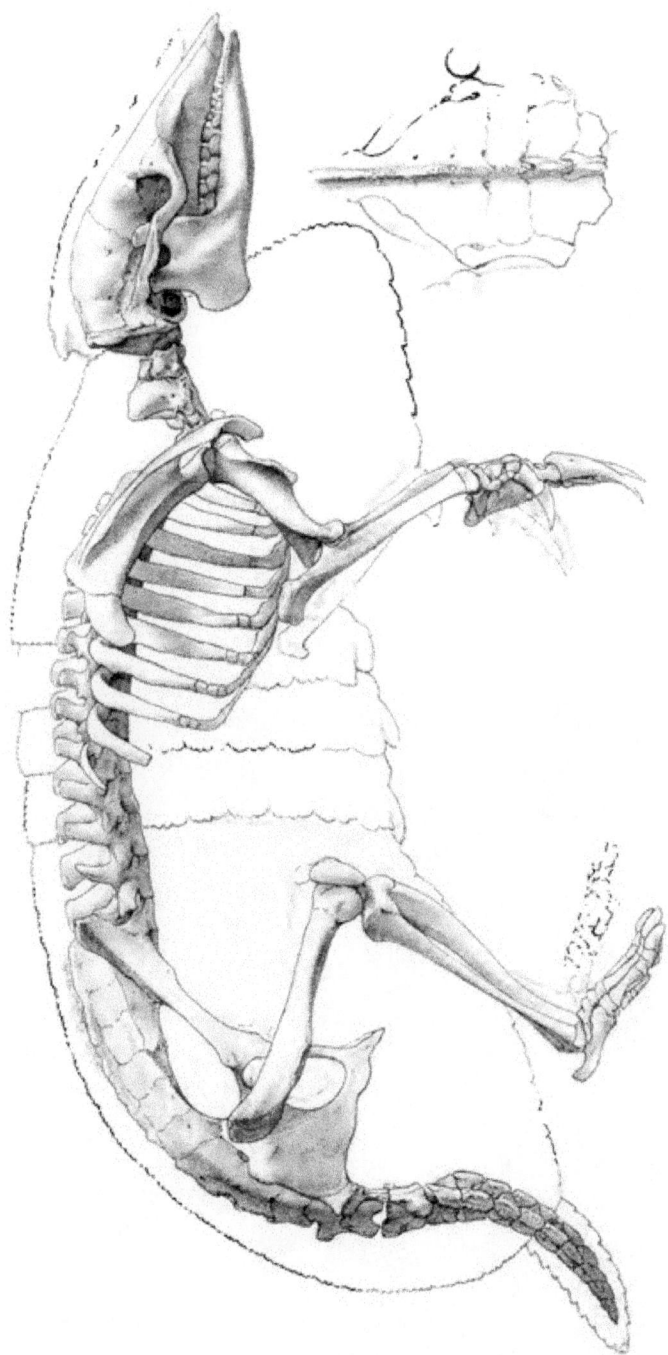

G.H.Ford & C.L. Griesbach.

Mintern Bros imp

1.

724.a.

2.

1535.C.

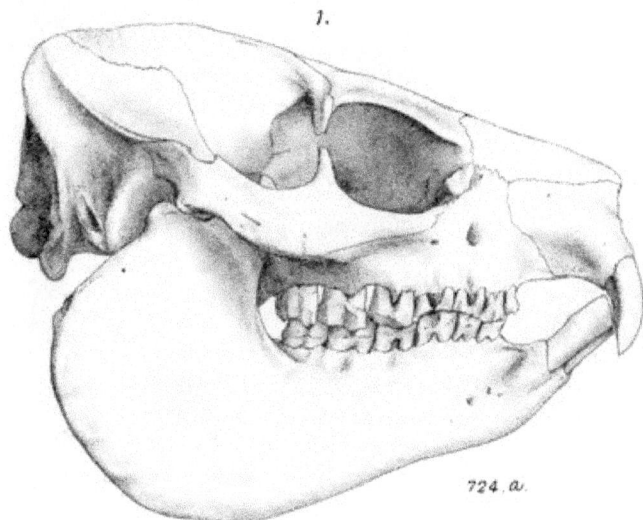

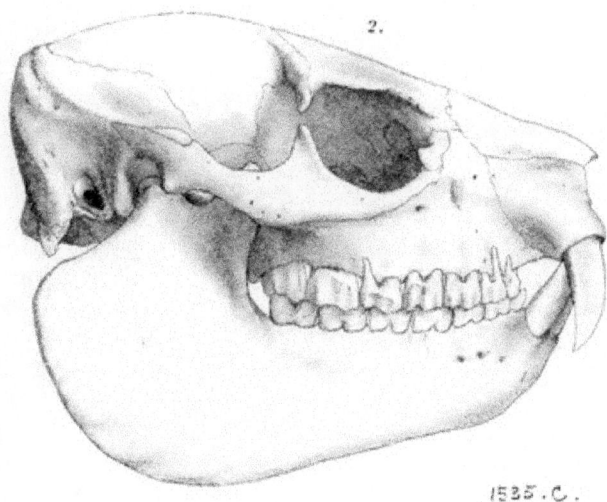

G H Ford & C L Griesbach.

Mintern Bros imp.

1. *Euhyrax abyssinicus*, p.43.
2. *Hyrax brucei*, p. 40.

1.

1391 *a*

2.

1515. *a*.

3.

58.5.4.
446.

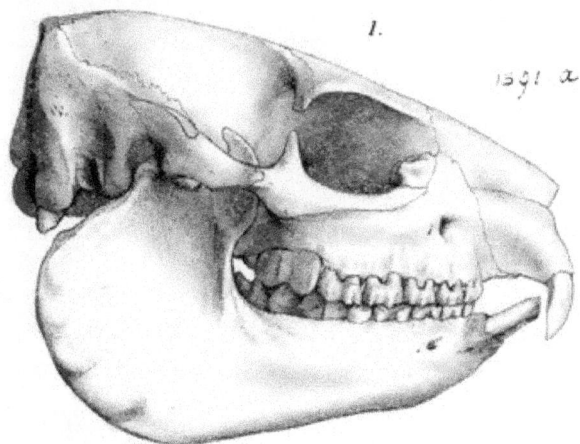

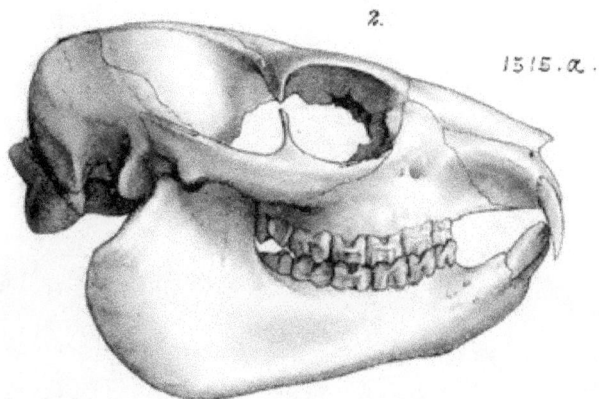

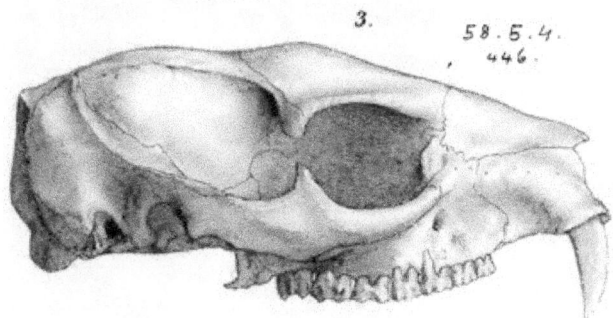

G.H.Ford & C.L.Griesbach.   Mintern Bros imp

1. *Hyrax ferrugineus*, p. 42.
2. *Euhyrax Bocagei*, p. 43.
3. *Dendrohyrax Blainvillei*, p. 44.

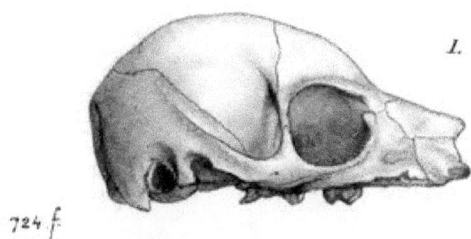

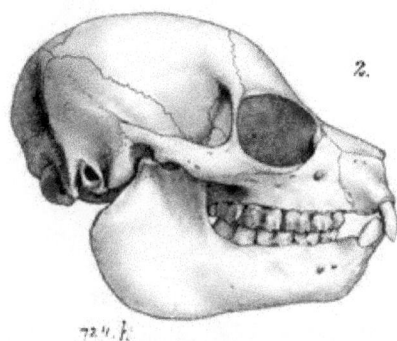

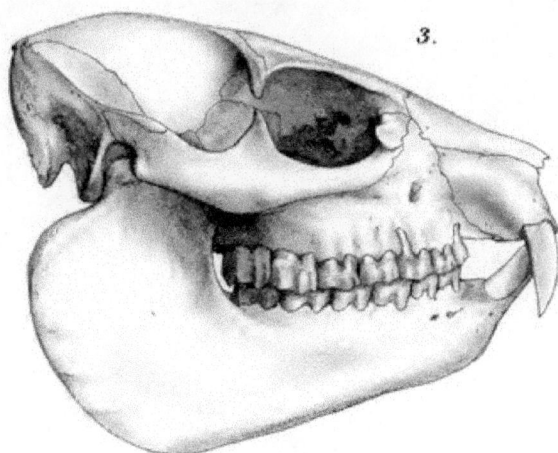

Pl. XII.

1.

724 f

2.

724 k

3.

1500. b.

G. H. Ford & C. L. Griesbach.

Mintern Bros. imp.

1. *Dendrohyrax dorsalis,* p. 43. very young.
2. „ *semicircularis,* p. 44.
3. *Hyrax irrorata,* p. 42.

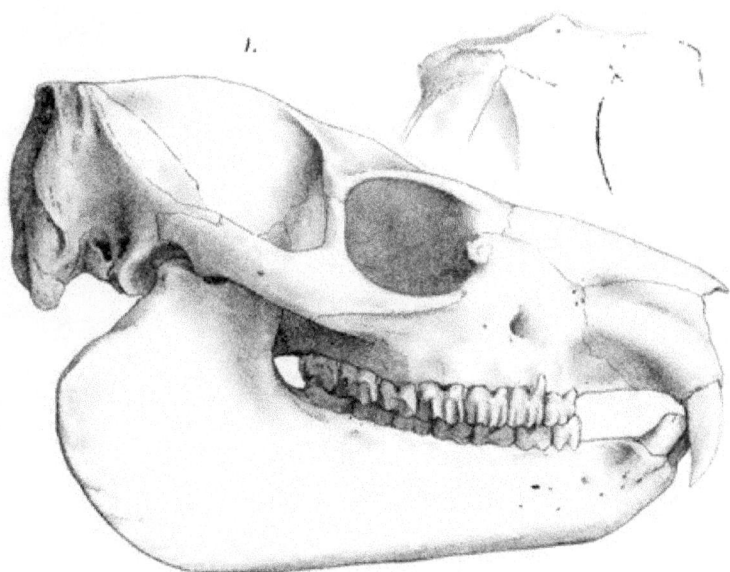

Pl. XIII.

1.

.1142. a.

2.

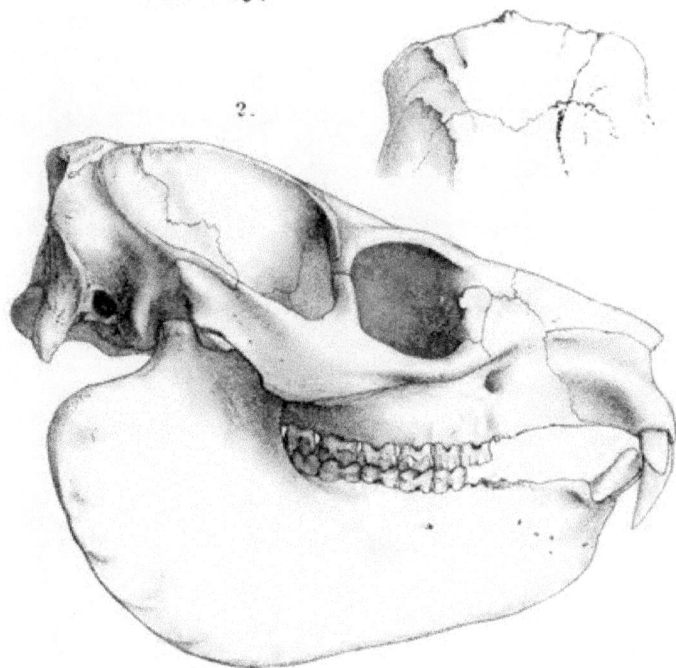

G.H.Ford & C.L.Griesbach.

1586. b.

Mintern Bros. imp.

1. *Dendrohyrax dorsalis*, p. 43.
2.   „   *arboreus*, p. 44.

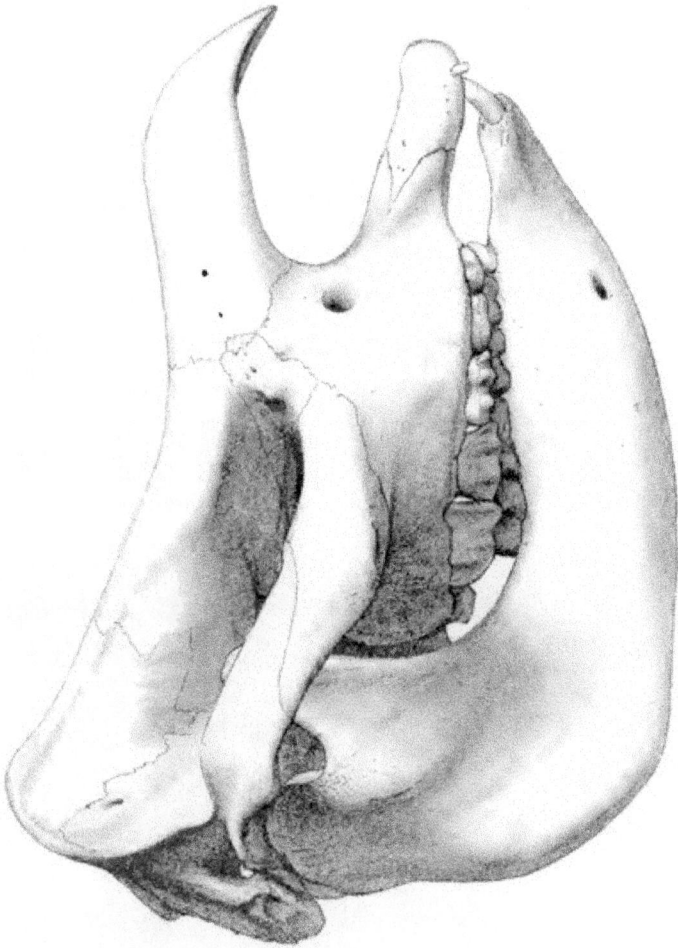

Pl. XIV.

Mintern Bros imp.

G.H.Ford & C.J. Griesbach

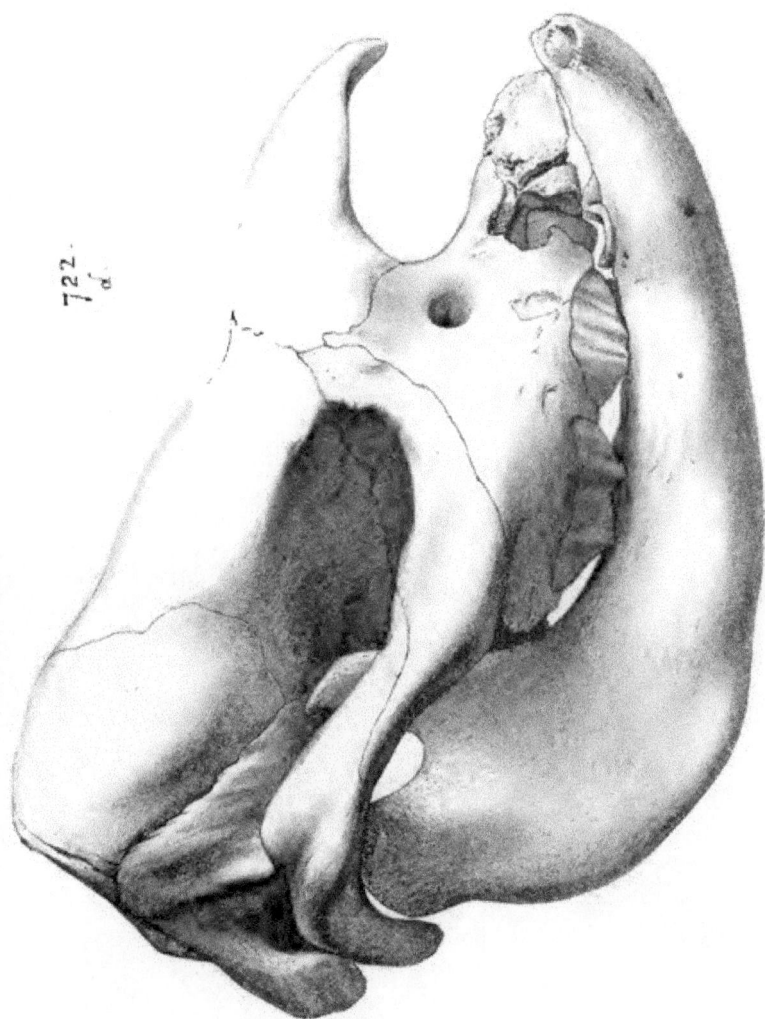

Pl. XV.

Mintern Bros. imp

722.
d.

G.H. Ford & C.J. Griesbach

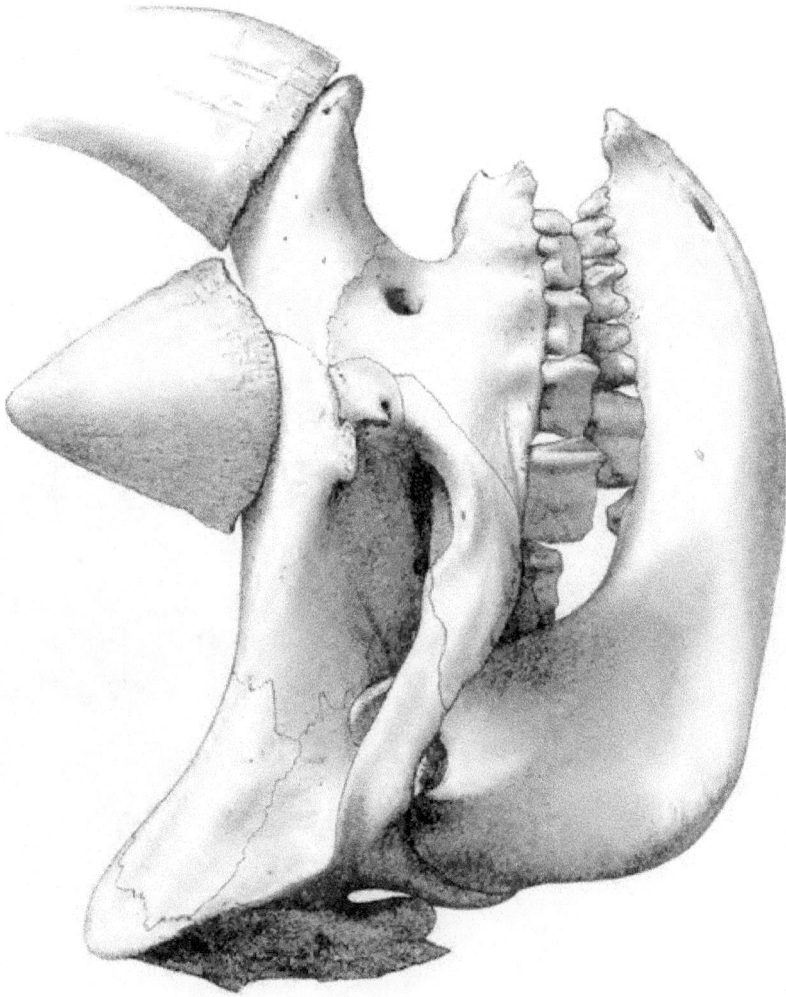

Pl. XVI.

Mintern Bros imp

G.H. Ford & C.I. Gnesbach.

Rhinaster bicornus. p. 51.

Pl. XVII

Rhinaster bicornis, p. 51. Abyssinia.

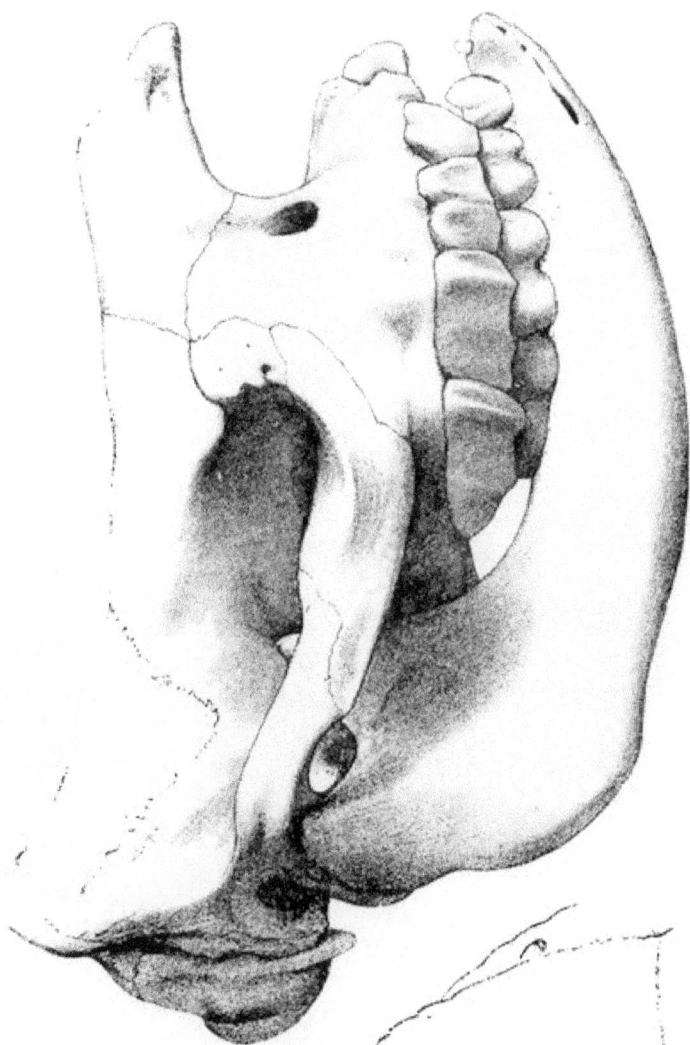

Pl. XVIII.

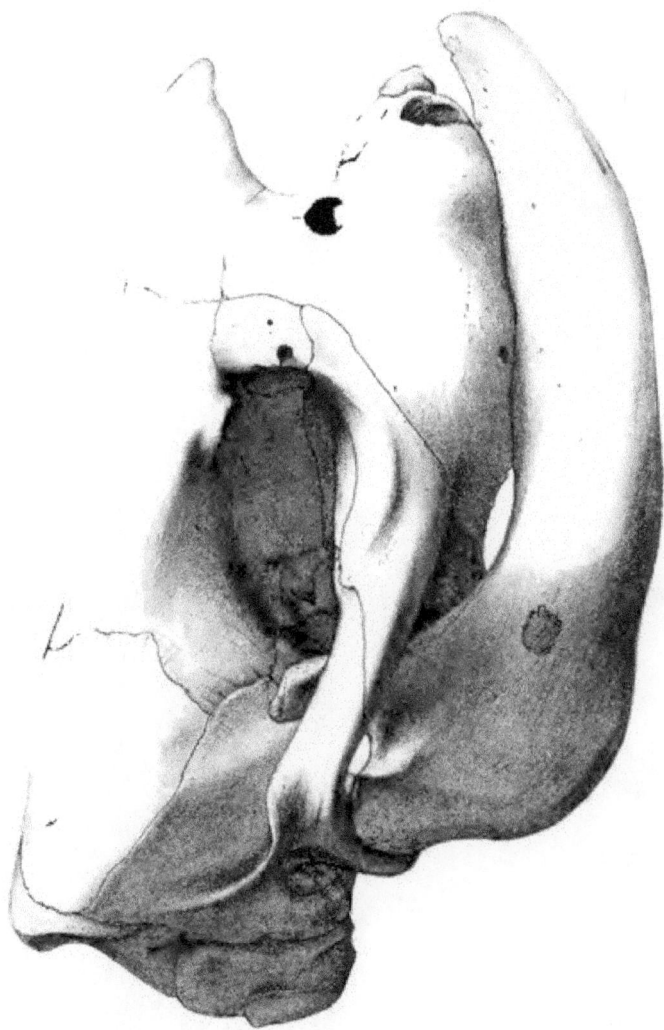

Rhinaster bicornis, p. 51. very young.

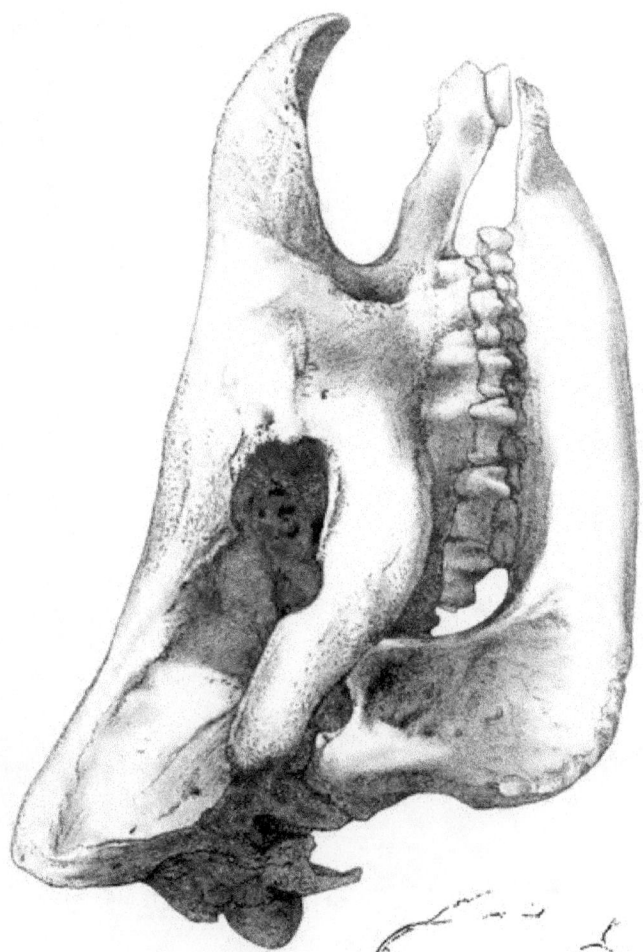

Pl. XXI

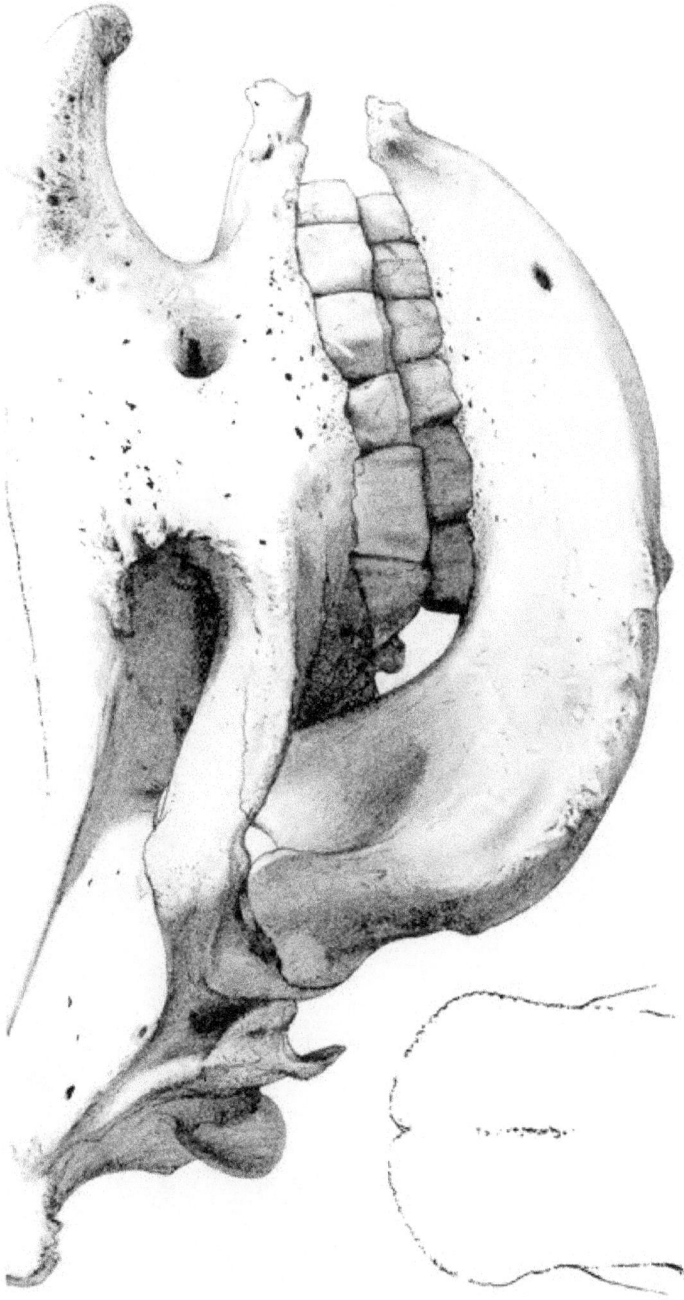

Hanhart Bros imp

Ceratotherium simum, p. 53.

G.H.Ford & C.L.Griesbach.

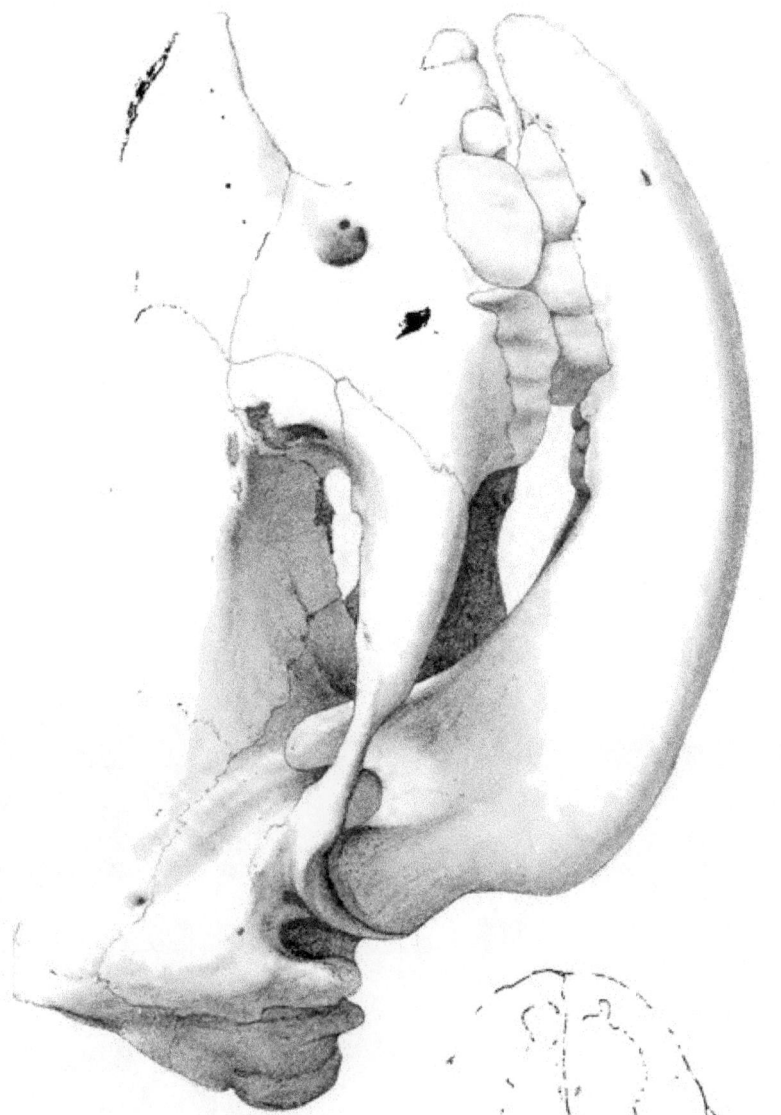

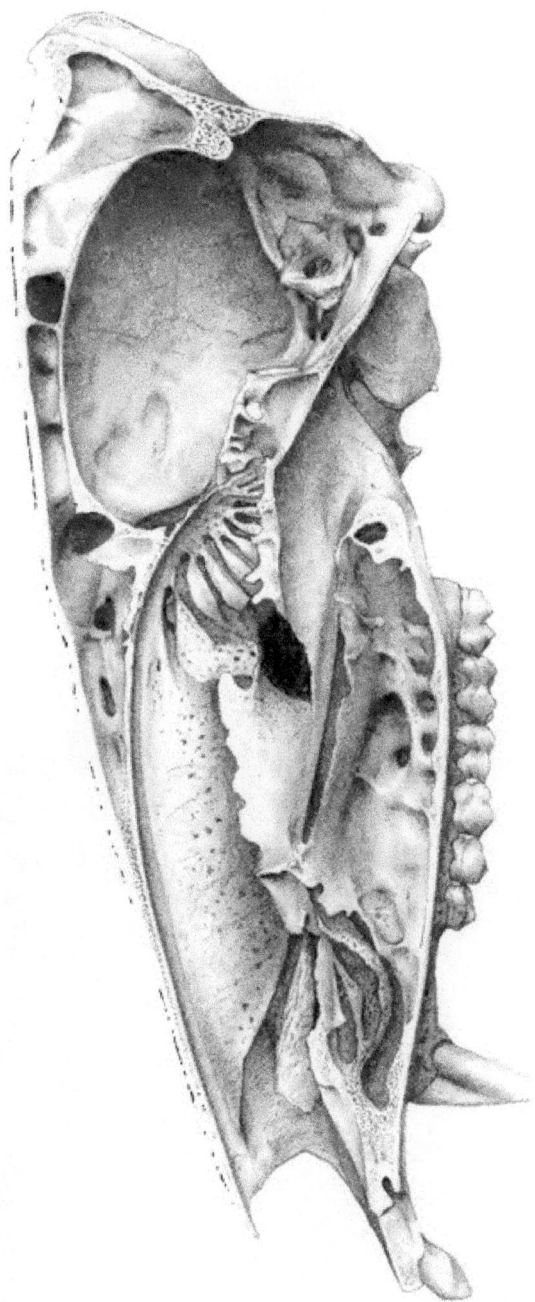

Pl · XXIII

1.

1363, a.

2.

1364, a.

3.

712, a.

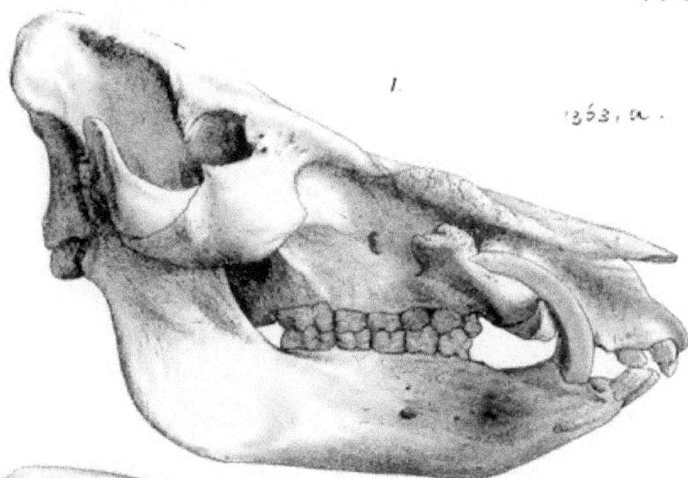

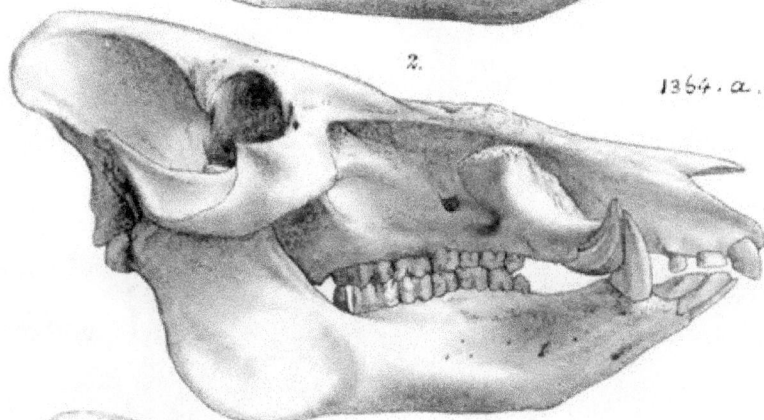

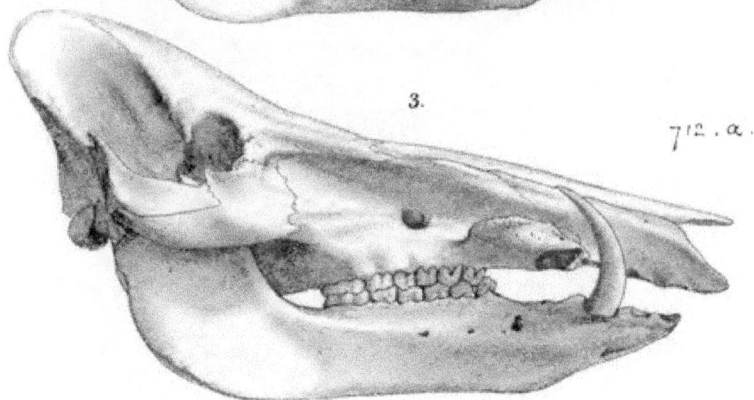

G. H. Ford & J. H. Griesbach   Mintern Bros. imp.

1. Potamochœrus porcus, p. 56.
2.   „   africanus, p. 56.
3. Euhys barbatus, p. 57.

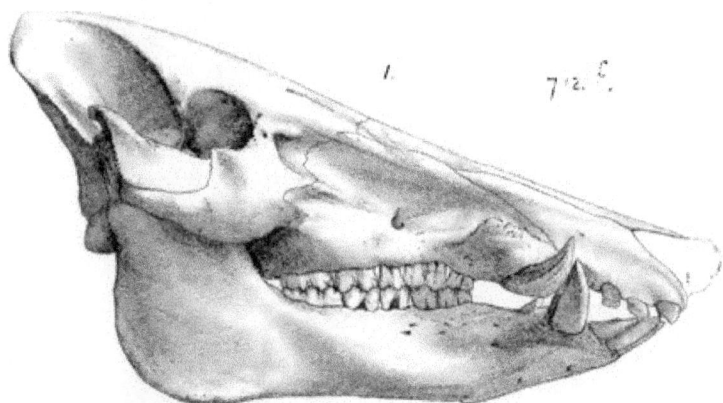

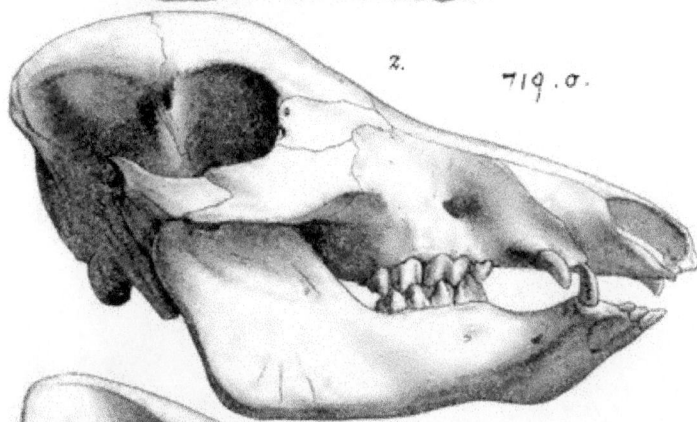

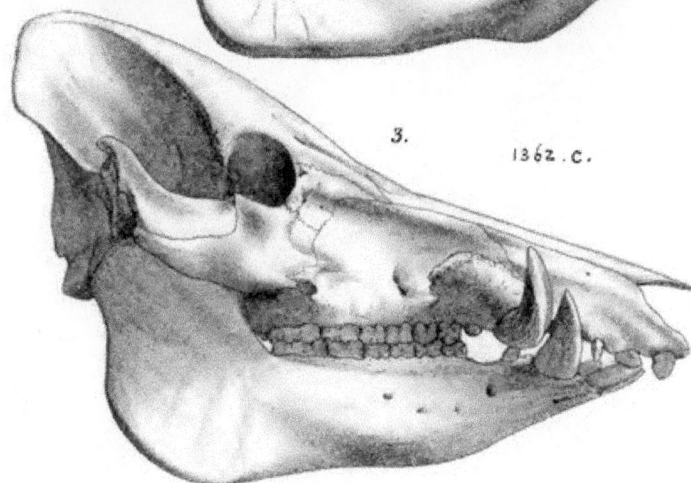

1.

712. ♀

2.

719. ♂

3.

1362. C.

1. *Dasychœrus verrucosus*, p. 59.
2. *Phacochœrus œthiopicus*, p. 69. *young*.
3. *Aulacochœrus villatus*, p. 58.

Pl. XXVII

1.   713.0.

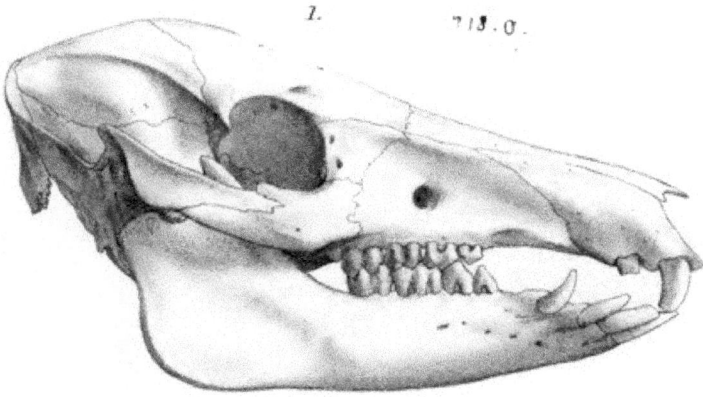

2.

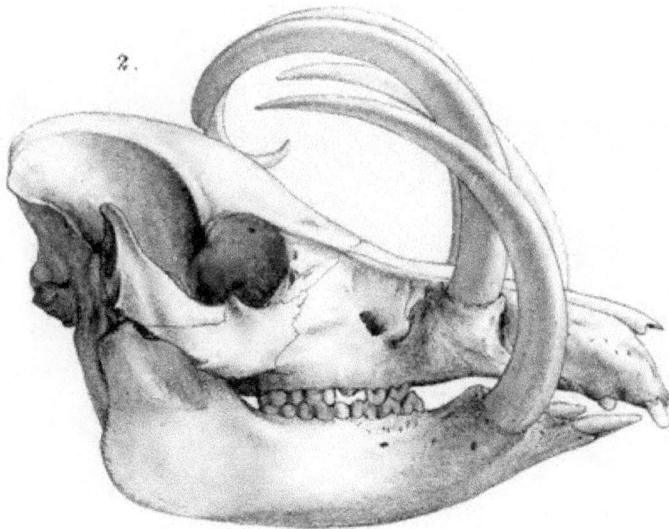

713.d.

G.H.Ford & C.L. Griesbach.　　　　　　　　　Mintern Bros. imp.

1. *Babirussa alfurus var.* ♀ p. 68.
2.   „   „   ♂ p. 68.

Pl. XXVIII.

1.

1457·a

2.

65597.

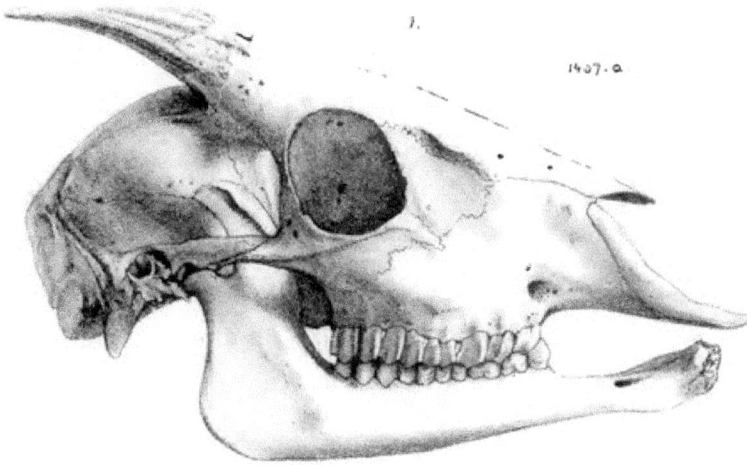

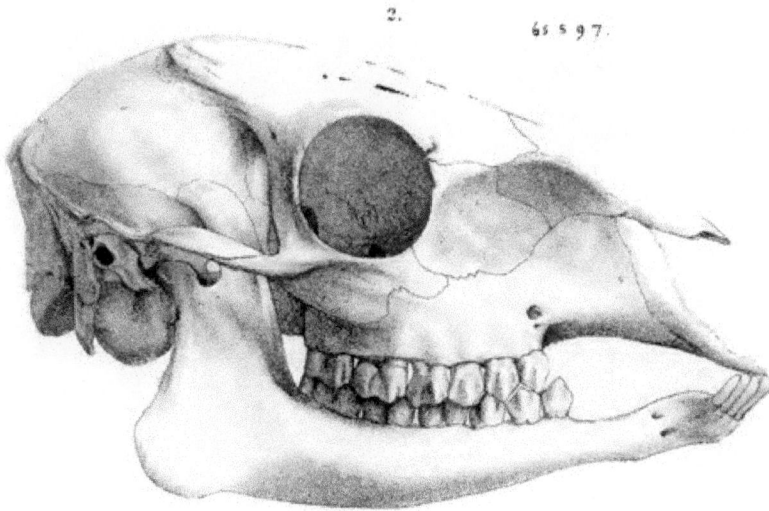

G.H Ford & C.I. Griesbach.

Mintern Bros imp.

1. *Capricornis swinhoei*, p. 92.
2. *Cephalophus nuhlatus adult*, p. 95.

95. c

1.

2.

1439. a

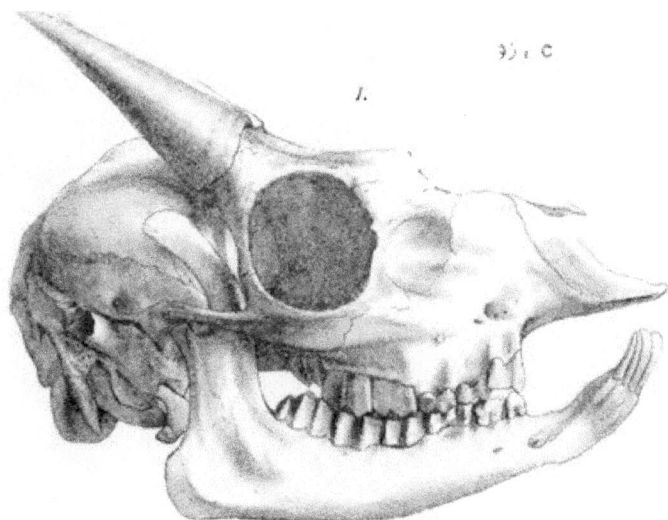

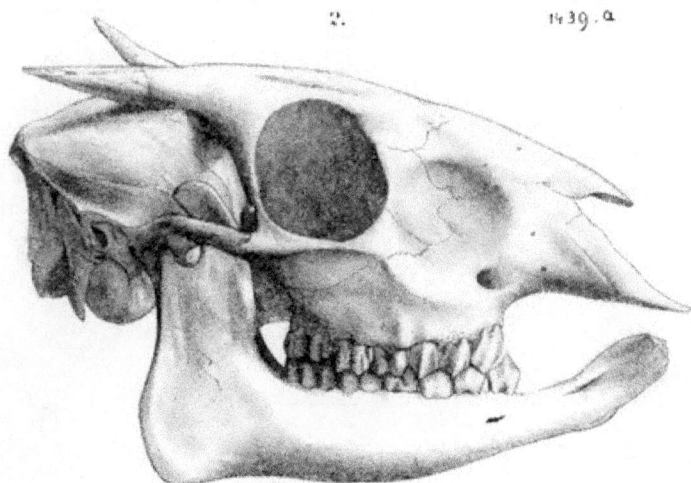

G.H Ford & C.L Griesbach

Mintern Bros imp

1. *Calotragus melanotis, p. 90.*
2. *Cephalophus dorsalis, p. 95.*

1.

71.7.3.7.

2.

624. ♭.

3.

1006. ♭.

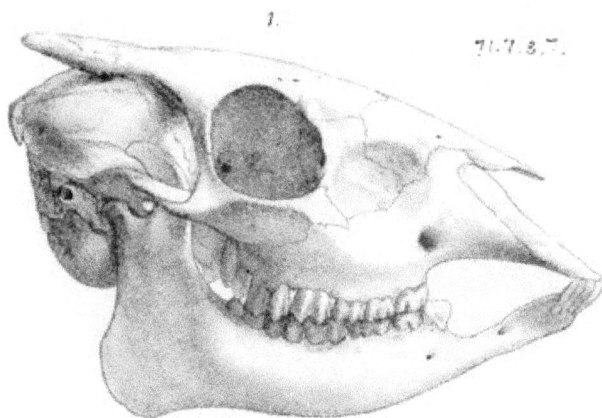

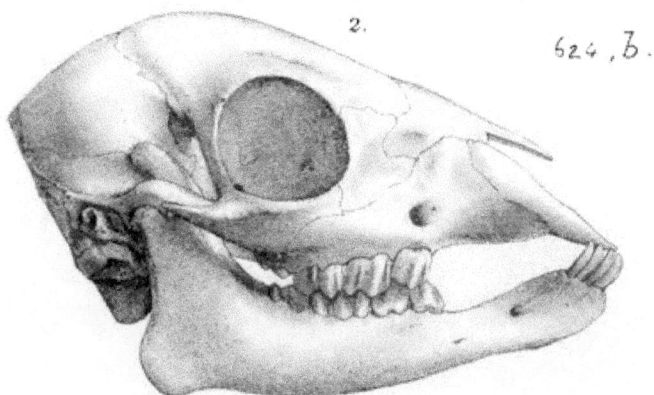

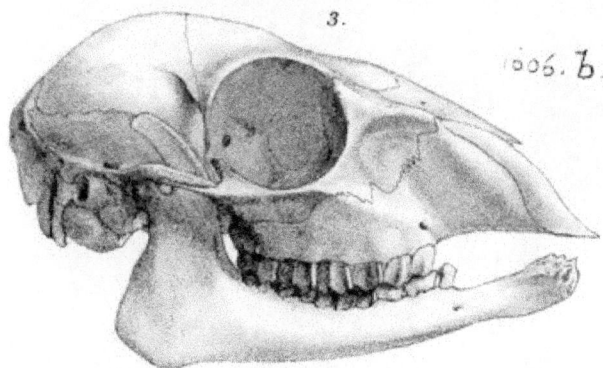

G.H.Ford & C.L.Greisbach.

Mintern Bros imp.

1. *Cephalophus badius, p. 94.*
2.   "   *coronatus, p. 96.*
3. *Nesobragus moschatus, ♀, p. 100.*

Pl. XXXI.

1.

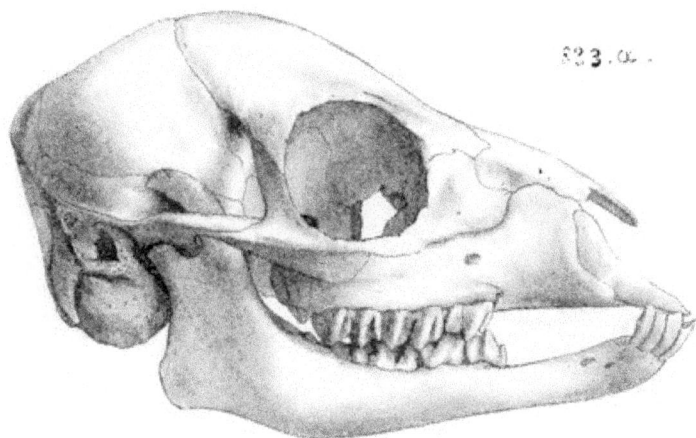

833. α.

2.

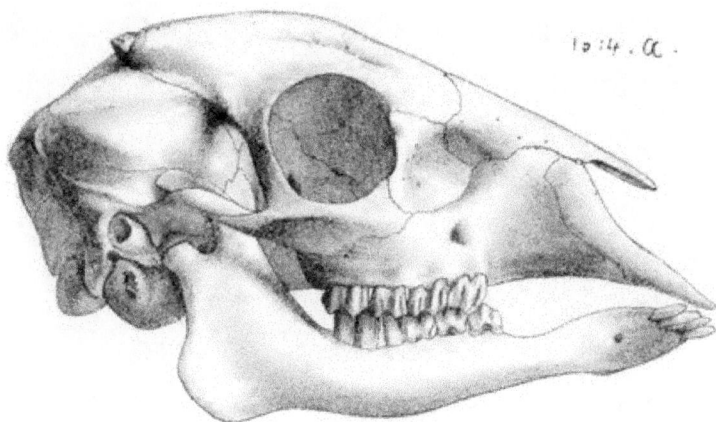

1914. α.

G. H. Ford & C. J. Griesbach.

Mintern Bros imp

1. *Cephalophus rufilatus, p, 95.*
2.   ,,   *aureus, p, 94.*

Pl. XXXII.

1.

2.

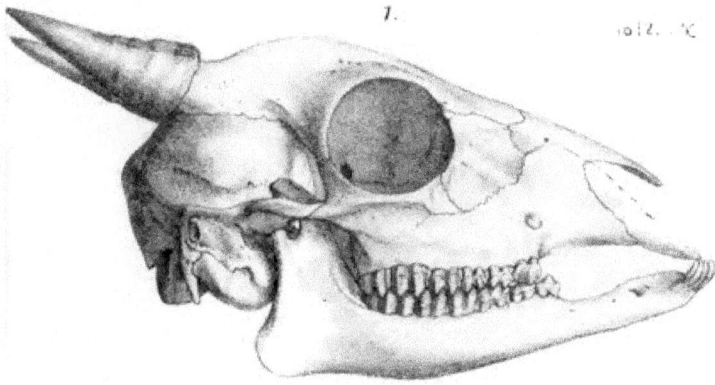

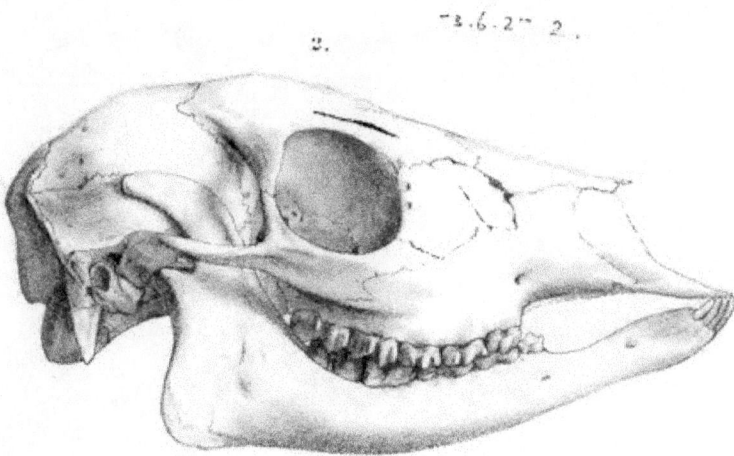

Mintern Bros. imp.

1. *Cephalophus melanorheus*, p.98.
2. *Cervus whitelyi*, p.

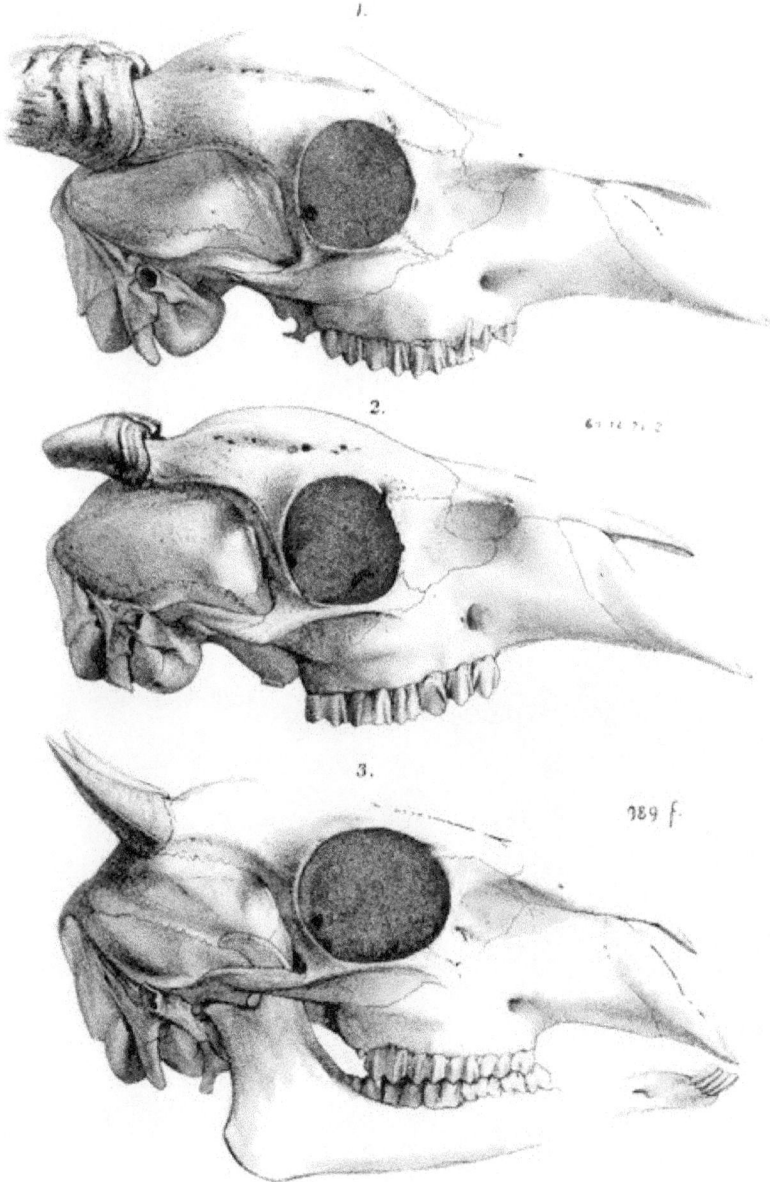

1. *Cephalophus ogilbyi*, p. 99.
2.   "   *ogilbyi*, p. 99.
3.   "   *pygmæus*, p. 97.

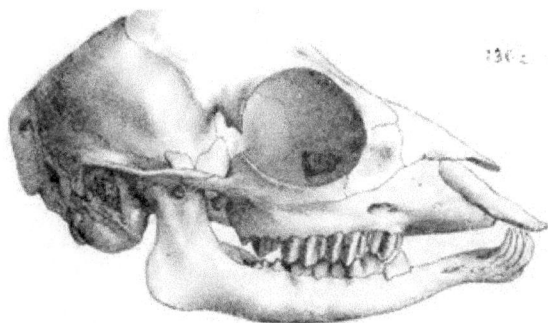

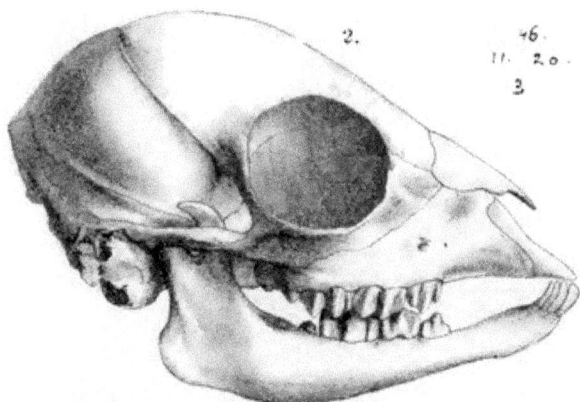

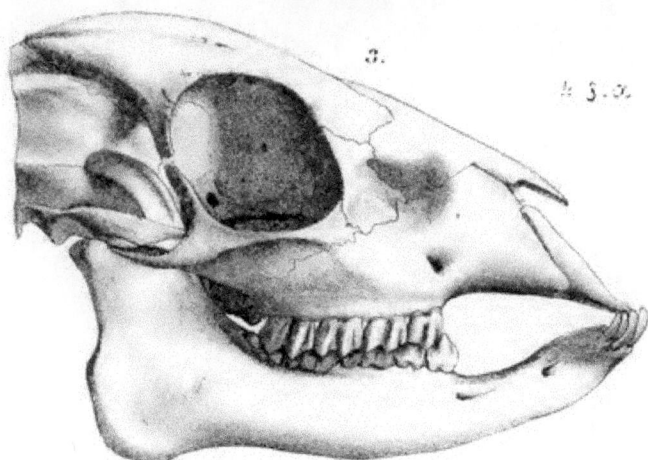

G.H.Ford & C.L.Griesbach.

Mintern. Bros. imp.

1. *Cephalophus bicolor*, p. 98.
2.   "   *whitfieldii*, p. 97.
3.   "   *badius*, p. 94.

Pl. XXXV.

1.

1375. α.

2.

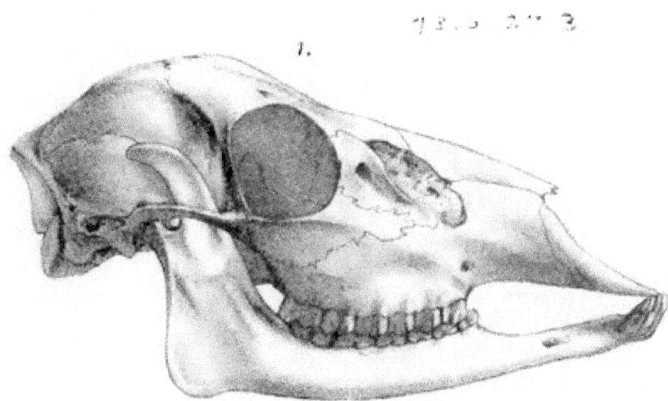

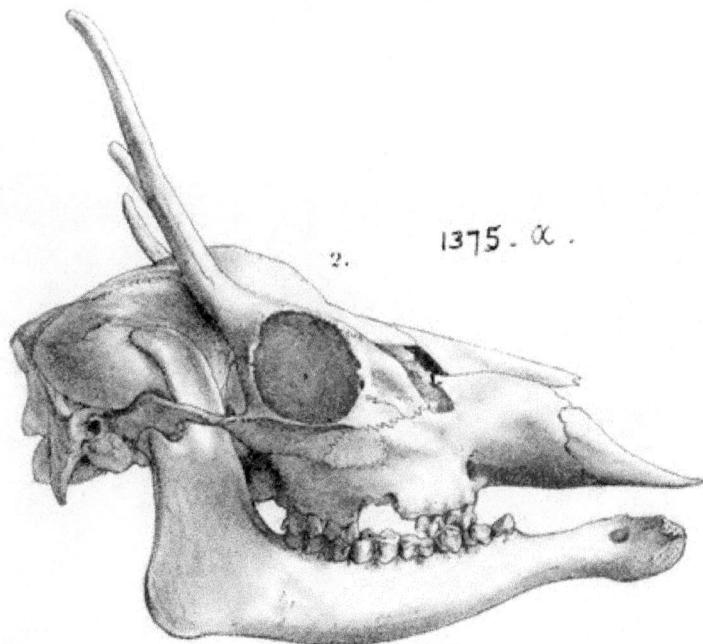

G.H.Ford & C.L.Griesbach.

Mintern Bros. imp.

1 & 2. *Furcifer antisiensis*, ♂ & ♀.

Pl. XXXVI.

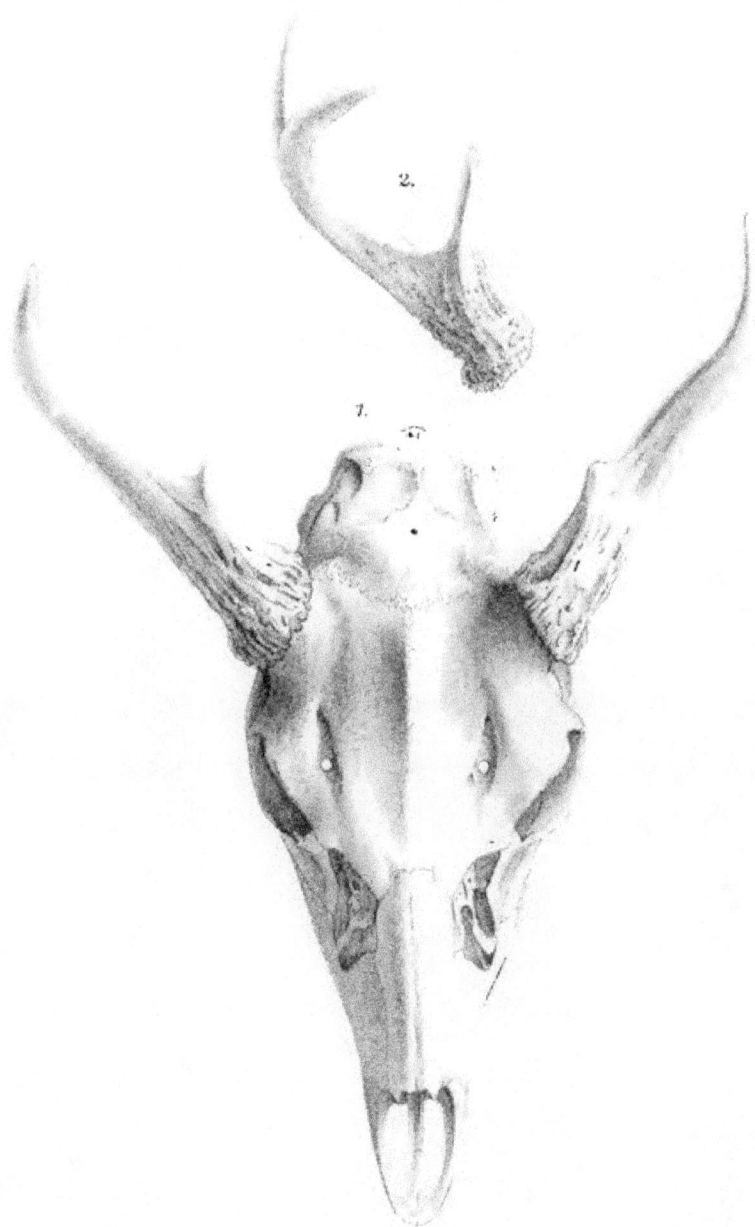

f.1. *Cariacus mexicanus.*
f.2. *Separate adult horn.*

Pl. XXXVII.

G.H.Ford & C.L.Griesbach.

Mintern Bros. imp.

Cariacus leptocephalus.

Pl. XXXVIII.

Eucervus macrotis.

Eucervus columbianus.

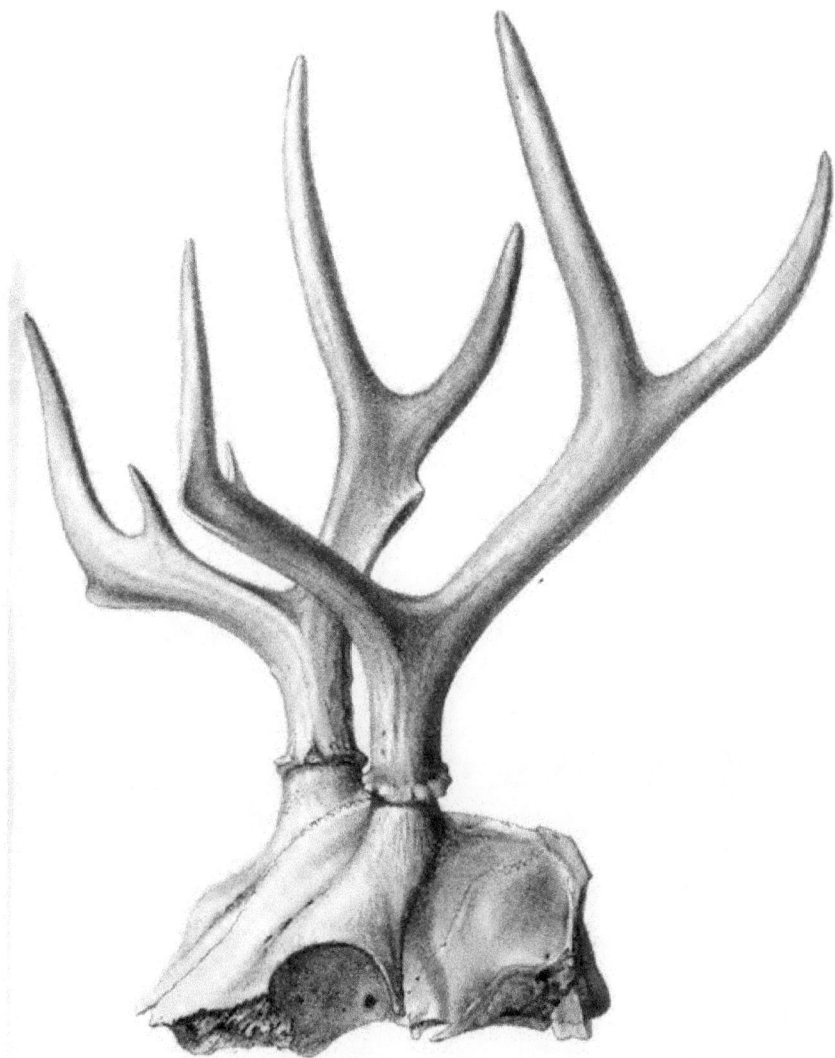

G.H.Ford & C.L.Griesbach.

Blastoceris sylvestris.

Mintern Bros.imp

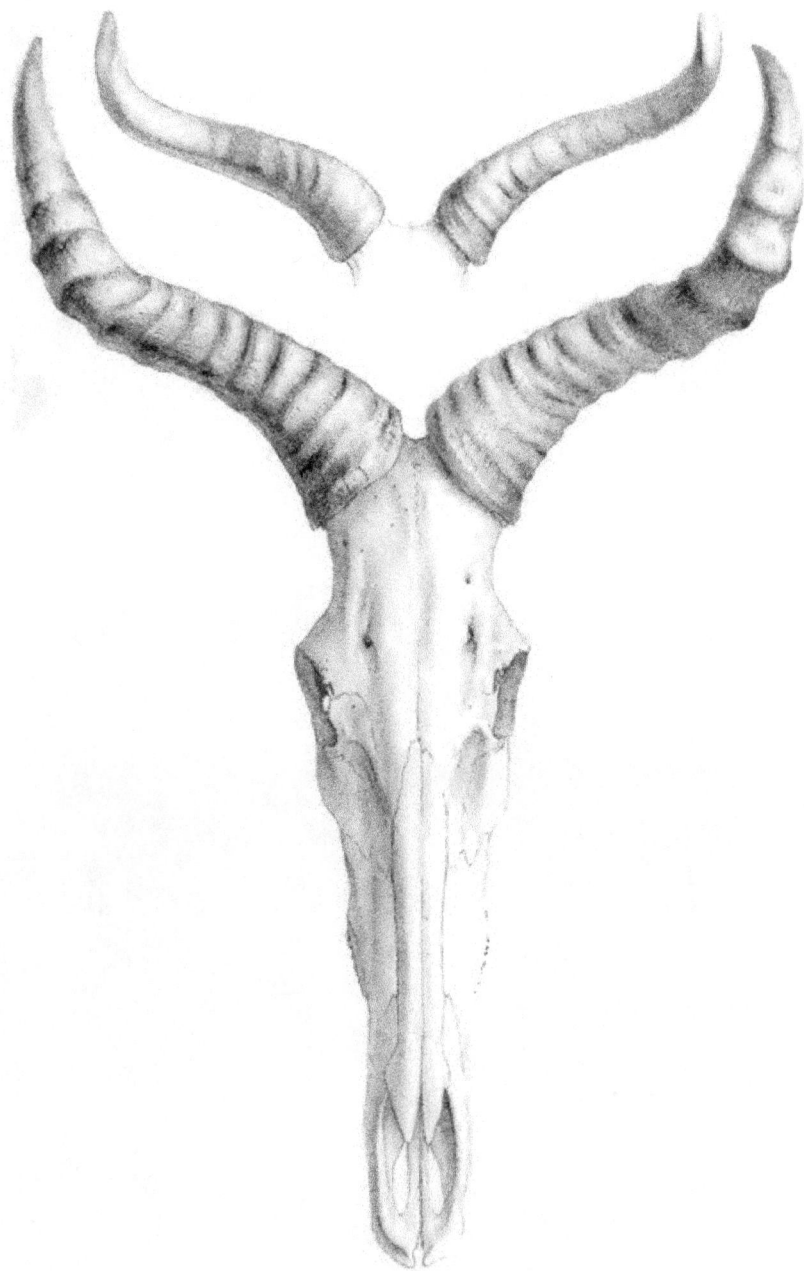

Alcelaphus tora.
Male & female.